Advanced Power Cable Technology

Volume I
Basic Concepts and Testing

Authors

Toshikatsu Tanaka, Ph.D.
Central Research Institute
of Electric Power Industry
Tokyo, Japan

Allan Greenwood, Ph.D.
Philip Sporn Professor
Director
Center for Electric Power Engineering
Rensselaer Polytechnic Institute
Troy, New York

CRC Press, Inc.
Boca Raton, Florida

Library of Congress Cataloging in Publication Data

Tanaka, Toshikatsu.
  Advanced power cable technology.

  Bibliography: p.
  Includes index.
  Contents: v. 1. Basic concepts and testing —
v. 2. Present and future.
  1. Electric cables. 2. Electric lines—Underground.
I. Greenwood, Allan, 1923—   . II. Title.

TK3351.T36 1983   621.319′2   82-14597
ISBN 0-8493-5165-0 (v. 1)
ISBN 0-8493-5166-9 (v. 2)

Direct all inquiries to CRC Press, Inc., 2000 Corporate Blvd., N.W., Boca Raton, Florida, 33431.

© 1983 by CRC Press, Inc.

International Standard Book Number 0-8493-5165-0 (v. 1)
International Standard Book Number 0-8493-5166-9 (v. 2)

Library of Congress Card Number 82-14597
Printed in the United States

PREFACE

We have learned much of past and present cable technology in the course of writing this book. An attempt has been made to describe the most recent information, but at the same time set out the fundamentals of cable technology in some depth. No details of "cable products" by individual manufacturers are described, instead, a focus is directed on general features of various kinds of cables. Although the expected prime readership will be cable engineers and those interested in cable research, the book will also be useful to university undergraduate and graduate students and management personnel who are interested in electric power engineering and underground power transmission in particular.

In the preparation of these two volumes abundant help was given by many distinguished cable research engineers in the U.S., Europe, and Japan. Mr. K. Masui of the International Electric Research Exchange helped in collecting related information internationally. We therefore accord him our sincere appreciation. Grateful thanks are also due to the following persons for their kind cooperation in obtaining necessary information and data: Dr. A. Lacoste of Electricite de France, Dr. P. Gazzana Prioroggia of Industrie Pirelli, S.P.A., Professor G. Wanser of Vorstansmitgelied der Kabel-und-Metallwerke Gute-hoffnungshutte Aktiengesellschaft, Dr. R. Patsch of AEG-Telefunken, Dr. J. D. Endacott of BICC Research and Engineering Ltd., Mr. R. Jocteur of SILEC, Mr. I. Eyraud of Les Cables de Lyon, Mr. R. W. Samm of EPRI, Mr. E. E. McIlveen of the Okonite Company, Mr. R. B. Blodget of the Anaconda Company, Mr. L. D. Blais of Kaiser Aluminum and Chemical Corporation, Dr. H. C. Doepkin, Jr. of Phelps Dodge Cable and Wire Company, the late Dr. Z. Croitoru of Electricite de France, Dr. C. M. Cooke of Massachusetts Institute of Technology, Dr. A. Cookson of Westinghouse Electric Corporation, and so many friends in the Japanese cable industry. In addition we wish to acknowledge the CRC staff for their incessant encouragement and elaborate editorial work. Finally, indebtedness is expressed to Hiroko Tanaka for her patience and encouragement and for her help in typing the manuscript. In this, at different stages, she was assisted by Amelia Stewart, Janice Daigle, and Jane Burhans, who must also share thanks.

Toshikatsu Tanaka
Allan Greenwood

THE AUTHORS

Dr. Toshikatsu Tanaka is a fellow research scientist at the Central Research Institute of Electric Power Industry in Tokyo, Japan. Dr. Tanaka received his B.A., M.S., and Ph.D. degrees in materials science from Osaka University in Osaka, Japan. He was Visiting Lecturer at the University of Salford, Salford, England from 1970 to 1972, and Associate Professor at Rensselaer Polytechnic Institute, Troy, New York, and Visiting Research Scientist at the General Electric R:D Center, Schenectady, New York from 1975 to 1976. Dr. Tanaka is a member of several engineering and scientific organizations and has been the recipient of an IEEJ (the Institute of Electrical Engineers of Japan) award for his scientific paper.

Dr. Tanaka has been engaged in electrical insulation studies for more than 15 years. He has published about 100 research papers including several reviews. He is also involved in computer application to the electrical insulation field and in new areas such as research projects of superconducting energy storage. He wrote one book in Japanese on the subject of high voltage and electrical insulation and is currently in charge of the electrical insulating materials section of an encyclopedia of electrical engineering to be published in the near future. He was formerly a contributer to one chapter of *Digest of Literatures on Dielectrics* by the National Research Council in the U.S. from 1971 to 1973.

Dr. Tanaka is also active in the IEEJ, IEEE, CIGRE, and IEC on electrical insulating materials, serving as chairman and secretary of certain committees of the IEEJ and as Convenor of CIGRE and IEC. He has served as a correspondent to *EI News in Japan* of the Electrical Insulating Society (EIS) Newsletter of IEEE since 1974 and was an AdCom Member of EIS from 1977 to 1979. Dr. Tanaka holds five Japanese patents.

Dr. Allan Greenwood is Philip Sporn Professor of Engineering at Rensselaer Polytechnic Institute, Troy, N.Y. where he is the Director of the Center for Electric Power Engineering. Dr. Greenwood received his early engineering education at Cambridge University (B.A. degree 1943, M.A. degree 1948) and his Ph.D. degree in 1952 from the University of Leeds. He has had an active career in industry, notably with the General Electric Co. His areas of expertise are power switching equipment, power system transients problems and power cables. He holds 16 U.S. patents and about 70 foreign patents, is the author of more than 50 scientific and technical papers (three of which received prizes), and the author or co-author of two books. He is a Fellow of IEEE and serves on the Fellow Committee of the Power Engineering Society. He is a member of CIGRE, acting as Convenor of its Working Group 13.03, and is a member of the honor society Sigma Xi. Dr. Greenwood consults for national and state governments and private industry.

TABLE OF CONTENTS

Volume I

Volume II

Chapter 1

INTRODUCTION

1.1. ELECTRIC ENERGY

Energy is an essential requirement for supporting life as we know it. Abundant, inexpensive energy has been an essential ingredient in the development of industrial societies. Without energy man would be at the mercy of his environment, his cities would be uninhabitable, and many of the material goods he now uses would be unavailable. Solar energy reaches the earth at the enormous rate of 170,000 billion kilowatts (BkW). It is ironic that at a time when new sources of energy are desperately needed, this vast resource is so difficult to harness. Approximately one half of the solar energy flux is converted into heat, about one fourth is reflected back to outer space, and the remaining fourth is consumed in generating winds and ocean currents. Photosynthesis accounts for about 40 BkW (0.023%) to create chlorophyll.

Compared with these astronomical figures, man's consumption of energy (10 BkW or 0.0058%) seems trivial, yet the expansion of this energy base is one of the most important issues in the world today. Although the U.S. has only 6% of the world population, it is responsible for about one third of the world energy consumption,[1] which is about 7×10^{15} btu (76 quadrillion btu, or 8×10^{19} J) annually as of 1975.[5] In the 1960s, the demand for energy was increasing by nearly 10% annually. The energy crisis of the early 1970s brought about a more moderate economic growth which was reflected in a drop in the annual energy growth rate to about 4%. The long-range forecast of energy production in the U.S. indicates a somewhat lower annual growth rate of around 3.5%.[5] There are some within the scientific community who believe that at some time in the future this growth rate will require the establishment of a ceiling on energy production. Since this heat must be absorbed by the earth and its surrounding atmosphere, there may be a limit to the amount of man-made waste heat which may be tolerated.

Figure 1.1. demonstrates the near-future prospect for energy demand[4] which covers a range of forecasts from the highest to the lowest growth rate. It is conservatively estimated that the U.S. energy demand will double in the next quarter century. This is smaller than past estimates and probably reflects a new assessment of the effects of energy conservation, energy price sensitivity, and a general acceptance of a slower growth in the economy.

Of particular interest in the context of this treatise is the fact that the proportion of the total energy supply being used to generate electricity is steadily increasing. This is probably because electricity is so clean and convenient to the end user. For many years growth in electrical energy consumption has been twice as great as the growth of total energy. The proportion of energy used for production of electricity increased from 11% in 1920 to over 20% in 1960. Moreover, this fraction exceeded 30% in 1980.

Energy forecasting is not a precise science, especially at a time when energy costs are increasing so rapidly. There is a built-in inertia in industrial societies where industrial progress has been traditionally linked with increased energy usage. Also, life styles which tend to be characterized by the profligate use of energy have become habitual and people are reluctant to change. On the other hand, industrial countries must now adjust to living under circumstances in which energy is expensive and some forms are in short supply. The manner in which government, industry, academia, and the public respond and adjust to this situation will be very important, and it will significantly influence the life style of society in the future.

If it is accepted, as many people believe, that U.S. electrical energy demand will quintuple in the next quarter century, it is apparent that some countries such as Japan and West

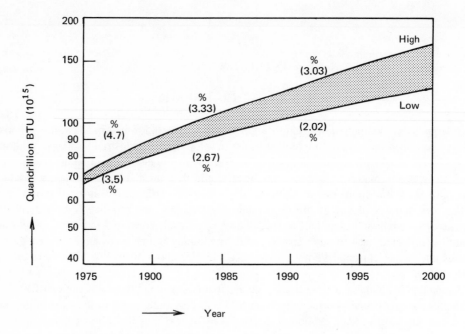

FIGURE 1.1.   Present forecasts of U.S. annual energy demand and growth rates.[4]

Germany will experience an even greater increase. On the other hand, other countries in Western Europe will probably experience a smaller increase. The figure will surely be extremely high in developing countries. There is continuing debate as to how the electrical portion of this demand will be supplied. Realistically, there are relatively few options. It is clear that oil must be conserved for purposes other than burning. It is equally clear that nuclear power is a quite viable option but it is far from acceptable in many sections of society. In some countries, the U.S. among them, there is a plentiful supply of coal, but burning coal presents many environmental problems. Much will depend upon the success of synthetic fuel developments based on coal. Natural gas remains an alternative for the near future but like oil, supplies are limited. It is likely that the effectiveness of nuclear fission will be substantially increased by breeder reactors. However, beyond this point the future of nuclear power through fusion is still in doubt. Nor do the prospects for alternative forms of energy, such as direct solar energy, wind and tidal energies, geothermal energy, or ocean thermal generation appear very encouraging on the massive scale required.

In summary, energy demand, world-wide, is increasing, albeit at a slower rate than formerly. The fraction of energy that ultimately is used in the electrical form is increasing at a considerably greater rate than energy usage as a whole, a trend that is likely to continue because of the flexibility and versatility of electrical energy. Regardless of how this energy is produced, this scenario inevitably leads to the conclusion that there will be a great need to expand the means to transport this energy from its point of generation to its point of delivery in the years ahead. This brings us squarely to the subject of this text.

1.2. UNDERGROUND TRANSMISSION

Almost invariably the center of load is not adjacent to the point of generation; the two are remote from each other, sometimes very remote. For example, hydropower generated at the dam on the Columbia River is transmitted to the city of Los Angeles. As we look to the future, we can expect to see the same pattern. The major coal resources in the U.S. are in states such as Wyoming and Utah, whereas the principal centers of load are on the east and west coasts or in such locations as Chicago.

There are basically two ways by which electricity is transported from generating plants to load areas, these are by overhead transmission lines and by underground cables. At the present time, the overhead option is heavily favored for economic reasons. In the U.S. only about 1% of the transmission lines are underground and most of these are in the major urban centers between Boston and Washington. As far as the economics are concerned, there are two factors that bring this situation about. The first is the cost of installation, which for cables is typically 6 to 20 times that of an equivalent overhead transmission line. The second relates to the thermal capacity of the respective circuits. When a cable is buried in the ground, it is to all intents and purposes placed in a thermal blanket, thus its transmission capability is thermally limited. This is not the case with most overhead lines which are cooled by convection in the ambient air. Rarely is the transmission capacity of an overhead line constrained by thermal considerations. Because of this it is often necessary, at a given voltage level, to put in several cables to transmit the same power as a single overhead line. For these reasons the use of underground cables has been largely limited to those places such as urban areas where overhead lines would be impossible because of safety or aesthetic factors, or simply that no right of way for such a line was available.

There are clear indications that the conditions just described are changing. This is a consequence of a number of factors, the first being the heightened awareness of the general public to the aesthetics of overhead transmission lines. As these have proliferated there has been a growing opposition to them for aesthetic reasons. This has been exacerbated as system voltages have risen because the higher voltages require taller towers with longer cross arms which are more prominent on the landscape. This has another implication which is the second factor militating against overhead transmission, namely, the amount of real estate that extra-high voltage (EHV) and ultra-high voltage (UHV) lines require. The intrinsic value of this land is no longer inconsequential. For example, an EHV overhead transmission line typically requires a 100-ft (30 m) wide right of way or about 12 acres/mi ($30,000 \text{ m}^2$/km); a UHV overhead transmission line needs more. If the installation cost of a 345 kV underground transmission line is $84,000/mi, this will be completely offset when the cost of land is $70,000/acre (in 1977 dollars). This may not be a significant factor in rural areas, but assuredly it is a factor in suburbia, and very important indeed in the urban situation.

Setting the cost aside, if transmission capability is to expand in the balance of the century to the degree anticipated, one may well ask whether land will be available for the necessary overhead transmission lines. Can we tolerate three or four transmission lines where now we have one?

Yet another consideration impacting on the mix of overhead and underground transmission circuits is the research and innovation being pursued to improve the position of underground transmission, particularly regarding its power transmission capability. Perhaps because of the much greater prevalence of overhead transmission lines, and perhaps also because of their relative simplicity, more attention has been paid to their optimization. At the present time and in the recent past, we are seeing an increasing effort to overcome the more important drawbacks of underground cable systems. This is concentrated in five basic areas:

A.   Reducing cost
B.   Simplifying installation and jointing
C.   Improving reliability
D.   Increasing the efficiency of power transmission
E.   Raising the transmission capacity of underground cable systems

As mentioned already, underground transmission is expensive compared with overhead transmission. This is unfortunate, but at the same time it presents a challenge. The higher cost is due in part to the cable itself and therefore encourages efforts to reduce the manufacturing cost, but installation is also an important item, being as much as 40% of the total cost. For this reason improvements in obstacle-detecting efforts, trenching, and tunneling techniques are being explored.

Cable installation, and especially splicing, requires considerable skill and is therefore expensive. Also, the kind of talent required for this type of work is in short supply. Attention is therefore being paid to devising means for simpler installation and jointing methods. Considerable progress has already been made for pre-prepared splicing kits for lower voltage cables. It is to be expected that this work will extend to higher voltage levels in due course. Past experience has shown that cable failures often occur at joints and terminations, thus, improvements in jointing techniques will enhance the overall reliability. Moreover, when failures do occur, these same techniques will reduce the cost of maintenance which is presently considerable.

The capacitance of underground cable is a very important consideration. On an AC system this must be charged and discharged every half cycle. The charging current involved is not inconsequential, indeed, in a present-day 345-kV cable, all the current-carrying capability of the conductor will be utilized for charging purposes when the cable is 42 km in length. This means that no useful current can be transmitted to a load beyond this distance. It is possible to compensate for the cable capacitance by the use of shunt reactors; however, these are expensive and occupy costly real estate. The longest underground circuits now in use in the voltage classes 230 to 345 kV are about 32 km in length.[9]

Cable capacitance is very dependent on the dielectric constant of the cable insulation. From this point of view, gas insulated cables are an attractive alternative. Unfortunately, they are more expensive.

Another dielectric property, the dissipation factor, assumes increased importance as the rated operating voltage rises. This impacts transmission capability insofar as it creates losses which raise the cable temperature. In this respect, all forms of losses are important. To raise transmission capability, means must be found to reduce the losses or to more effectively remove the heat that they give rise to. From the point of view of dielectric losses, DC operation is attractive; also, charging current is no longer a problem with DC. Advanced technologies such as cryoresistive and superconducting cables can similarly reduce the ohmic losses. Various forms of forced cooling can diminish the effects of such losses as are generated. By these methods, individually or in combination, it should be possible to attain power ratings for individual cable circuits of 2 GW or more.

Electrical insulation is a determining factor in cable design; for this reason cables are usually named for their insulation system. There are basically three kinds of cable insulation: tape insulation, solid insulation, and a gas insulation. The principal tape insulation is oil-impregnated cellulose paper. This paper is lapped onto the cable conductor and then impregnated with oil; thereafter it may or may not be pressurized. The pressurized type represents a mature technology that has been extensively developed and presently represents the most stable and probably the best existing insulating system. It exists in two categories, self-contained oil-filled (OF) cables and pipe-type OF (POF) cables. The bulk of the underground cable systems installed in the U.S. are high-pressure, oil-impregnated paper-insulated, oil-filled, pipe-type cables. In contrast, self-contained OF cables are a common practice in Europe and Japan.

Solid insulation is usually extruded onto the conductor. Many different materials are used, they include polyethylene, cross-linked polyethylene, butyl rubber, ethylene-propylene co-polymer, or ethylene-propylene-dien terpolymer. Since 1970 an increasing amount of ex-truded, chemically cross linked, polyethylene cable has been installed, initially in the voltage range between 46 and 69 kV, but more recently at higher voltages between 115 and 138 kV. Cross-linked polyethylene-insulated cables rated 230 kV and 345 kV are now under development. A primary objective in this development is to increase the working stress by dramatically reducing harmful imperfections in the insulation.

Table 1.2.
UNDERGROUND CABLE SYSTEMS MVA CAPABILITY[8]

	MVA Rating
High pressure, oil-impregnated paper-insulated, oil-filled pipe cables (HPOF)	
Selfcooled (550 kV)	950
Forced-cooled (765 kV, $T_s = 40°C$)	2,000
High pressure, oil-impregnated synthetic-insulated, oil-filled pipe cables (HPOF)	
Selfcooled (765 kV)	1,600
Forced-cooled (765 kV, $T_s = 40°C$)	2,600
Extruded cross linked polyethylene-insulated cables (XLP)	
Directly buried (345 kV)	800
DC cables oil-impregnated paper-insulated, oil-filled cables	
Self-cooled (± 600 kV)	2,500
Cryogenic cables	
500 kV	5,000
Superconducting cables	5,000—10,000
Gas insulated cables	5,000

When we speak of gas insulation we are usually referring to compressed SF_6 gas. Cables with this type of insulation would appear to be most appealing for long distances because the dielectric constant is minimal and the dielectric dissipation factor is virtually zero. However, at the present time this insulation system is used only for getaways from substations. The problem in other areas is cost, which is therefore a principal concern in the research and development work being done on such systems.

Cryogenic resistive cables and, perhaps more important, superconducting cables appear to hold considerable promise for the long term. Considerable progress has been made in developing both of these technologies, though some problems still remain. Vacuum insulation and cellulose or synthetic papers impregnated with liquid nitrogen appear to be candidates for the insulation system of cryoresistive cables. For superconducting cables, vacuum insulation and helium- (liquid or supercritical) impregnated polymer film-lapped insulation is presently under consideration. As things presently stand, it is doubtful whether such cables would be used if they were commercially available. This is because utility companies are as yet unprepared to commit to a single circuit the power that would make such a cable system attractive. However, as load grows and transmission systems expand, this situation may well change.

Table 1.2. gives a summary of the MVA capabilities of various underground cable systems.[8] The transmission power limits for OF and POF cables are controversial, but are probably less than 2 GW even when forced cooling is used. Cryogenic resistive cables and compressed gas-insulated cables, on the other hand, could probably economically transmit powers exceeding 2 GW. Superconducting cables will probably be used only where it is required to transfer 5 GW or more, though smaller cables, perhaps 2 GW, may be installed in special cases. DC cables appear to be the only solution for long-distance, high-powered underground or underwater transmission systems. Present thinking indicates that oil-impregnated paper is the soundest and most appropriate insulation system for this application.

This book presents up-to-date information on advanced cable technologies for both conventional and future underground power transmission systems, (Figure 1.2.) with particular emphasis on bulk power transmission and the performance of the electrical insulation of the respective cables. Chapter 2 of this volume stresses the importance of material science in cable design, especially the properties of the dielectrics used. It contains an abundance of

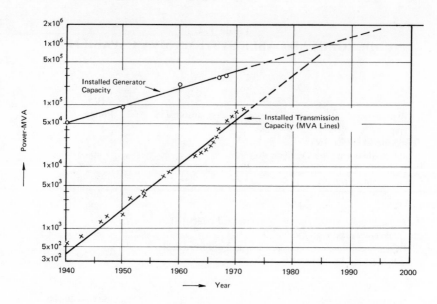

FIGURE 1.2.   Comparison between generation and transmission capacities.[8]

material applicable to cable design. Chapter 3 (Volume I) reviews testing, maintenance, and fault location procedures for conventional cables and recent developments in partial discharge measuring techniques and aids to the investigation of cable defects. Chapter 1 of Volume II introduces the reader to the state of the art in the many power cable technologies presently available; it goes into the details of construction and, in many instances, the manufacturing processes involved. This chapter also indicates where work is in progress to improve cable performance and indicates trends to be expected in the future. Chapter 2 (Volume II) is concerned with completely new types of underground cables which are presently under development and describes recent advances and prospects for the future.

REFERENCES

1. **Hammond, A. L., Metz, W. D., and Maugh, T. H., II,** *Energy and the Future,* Monograph, American Association for the Advancement of Science, Washington, D.C., 1972.
2. **Abelson, P. H.,** *Energy: Use, Conservation and Supply,* Monograph, American Association for the Advancement of Science, Washington, D.C., 1974.
3. **Felix, F.,** Point-of-use efficiency gives electric energy an edge in generating high per-capita GNP, *Electr. World,* November 1975, p.64.
4. **Hauser, L. G.,** Future Trends in Energy Supply, Electr. Utility Eng. Conf., Subject 22, Washington, D.C., March 28—April 9, 1976, 1.
5. **R & D Goals Task Force to ERC,** *Electric Utilities Industry R & D Goals Through the Year 2,000,* ERC Publ. No. 1—71, Electric Research Council, New York, June 1971, 1.
6. **Iveson, R. H., LeBlanc, D. J., and Garver, L. L.,** Planning transmission for the year 2,000, *Transmission Distribution,* September 1975, p.32.
7. **Corry, A. F. and Kasum, E.,** Future transmission-underground?, *Proc. Papers of 1972 IEEE Underground Transmission Conf.,* May 1972, p.1., IEEE, Piscataway, N.J.
8. **Bahder, G. and Eager, G. S.,** *Reviews of Present and Future Underground Transmission Systems,* APPA Eng. and Operation Workshop, American Public Power Assoc., Washington, D.C., February 1976, 1.
9. **Avila, C. F. and Corry, A. F.,** Underground transmission in the United States, *IEEE Spectrum,* March 1970, p. 42.

Chapter 2

MATERIALS SCIENCE IN CABLE TECHNOLOGY

2.1. CONCEPT OF CABLE DESIGN

2.1.1. Cable Structure[1-7]

The power cable to be discussed in this book is intended to transmit electric energy or electric power. Such cables are usually buried underground. They are therefore called underground residential distribution lines or underground power transmission lines according to their functions. The cable is an insulated conductor. Power cables or simply cables are classified according to their type of insulation as paper, rubber, plastic, or gas. Cables for power transmission and distribution are composed of many different types of insulation, conductors, and sheathing materials. Joints and terminations are the other important parts of cables or cable systems.

A conceptual design of a cable is shown in cross section in Figure 2.1.1. It will be observed to consist of two basic components, the conductor and the insulation. The conductor and insulation shields and metallic sheath are of considerable importance. Two other additional components, the sheath for corrosion protection and mechanically reinforcing metallic belt or wire complete the design. Power cables are classified according to their material structure as shown in Table 2.1.1. Sheaths may be bonded to minimize sheath losses and sheath voltages.

2.1.2. Electrical Design[1-7]

A. Insulation and Shield

Insulation thickness of a power cable is determined by recognizing that it must withstand not only the steady state AC voltage but also transient lightning impulses and switching surge voltages. As a result of many years of testing, it has been established that the performance of the insulation is principally determined by the maximum electrical stress in the insulation. This is especially true for oil-impregnated paper-insulated cables. The maximum stress occurs in the insulation immediately adjacent to the conductor in a cable consisting of coaxial cylindrical electrodes. Assuming a uniform permittivity it is given by the following:

$$E_{max} = \frac{V}{r \ell n \, (R/r)} \qquad (2.1.1.)$$

where V = applied voltage, r = radius of the inner electrode, and R = radius of the outer electrode.

It would be preferable electrically if the conductor were a smooth cylindrical rod or tube with a constant curvature. Unfortunately, rods and tubes have little flexibility and, since the cable must be coiled, a stranded conductor is used which has a far from smooth surface.

A smooth conductor surface is obtained by lapping the conductor with several layers of metalized paper, carbon paper, or metalized carbon paper and Duplex® paper for taped cable such as OF cable and POF cable. For extruded cable, it is achieved by lapping the conductor with carbon-filled cloth or nylon tape or by extruding over the conductor a covering of ethylvinyl acetate (EVA) carbon-filled copolymer, polyethylene, or cross-linked polyethylene.

Carbon paper has the advantage of conductivity through the paper; a conductor screen can be formed with two layers. Certain other advantages are claimed for its use, but one disadvantage is that particle activity at the surface facing the dielectric causes a rise in power factor with voltage, which can hide the presence of harmful ionization. Metalized carbon

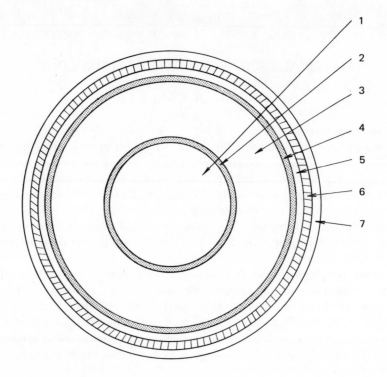

1. Conductor
2. Conductor Shield
3. Insulation
4. Insulation Shield
5. Metal Sheath
6. Corrosion-Protected Sheath
7. Metal Belt or Wire Armor

6. and 7. are subject to change according to the type of cable.

FIGURE 2.1.1.   Basic cross-sectional structure of power cable.

paper has the same resistivity (10^5 to 10^8 Ω-cm) as carbon paper, but presents a metal surface to the dielectric so that there is little, if any, incremental power factor effect. Semiconducting EVA copolymer, polyethylene (PE), or cross-linked polyethylene (XLPE) are preferable to semiconducting tape for the higher voltage PE or XLPE extruded cable because of their more stable performance. The volume resistivity of the semiconducting compound should be less than 10^4 to 10^5 Ω-cm.

The conductor shield is provided to prevent any corona discharge between the conductor and the insulation and therefore is required to make good contact with the insulation. Intersurface cross linking is preferable for XLPE cable. Two incompatible requirements are forced on the insulation shield. The first is similar to the requirement for the conductor shield, but the second requires that the insulation shield be easily stripped in termination and joining operations. The stripability may be controlled for cross-linked type semiconducting shields by controlling the degree of interface cross linking. This results in what is called a free-stripping semiconducting shield for extruded cable.

Extruded insulated power cable is normally produced with a metallic shield composed of helically applied copper tapes. When the cables are used on transmission circuits, this shield is supplemented with copper wires or ribbons to provide adequate short circuit capability. Recently, an improved metallic shield has been introduced.[37] This shield consists of a

Table 2.1.1.
CLASSIFICATION OF POWER CABLES AND THEIR EXAMPLES

Belted paper cable
  Belted paper-insulated lead-sheathed cable
  Belted paper-insulated lead-sheathed chloroprene-corrosion-protected cable
  Belted paper-insulated lead-sheathed jute-wound cable
  Belted paper-insulated lead-sheathed steel-belt-armored chloroprene-sheathed cable
  Vertical riser belted paper lead-sheathed double iron wire-armored cable
  Under-water-use belted paper lead-sheathed single iron wire-armored cable
  Belted paper insulated aluminum-sheathed polyethylene-corrosion-protected cable
    Conductor — compacted conductor, segmented compacted conductor

SL or H type paper cable
  SL paper-insulated chloroprene-sheathed cable
  SL paper-insulated steel-belt-armored cable
  SL paper-insulated chloroprene-corrosion-protected steel-belt-armored cable
  SL paper-insulated vinyl-sheathed cable
  H-type paper-insulated lead-sheathed cable
  H-type paper-insulated chloroprene-corrosion-protected cable
    Conductor — compacted conductor, aluminum-sheathed-type available

Gas-pressure cable
  Low-pressure gas-filled paper-insulated lead-sheathed chloroprene-corrosion-protected cable
  Low-pressure gas-filled paper-insulated lead-sheathed chloroprene-corrosion-protected steel-belt-armored cable
  Low-pressure gas-filled paper-insulated aluminum-sheathed vinyl-corrosion-protected cable
    Conductor — compacted conductor, segmented compacted conductor

OF cable (oil-filled cable)
  OF paper-insulated lead-sheathed chloroprene-corrosion-protected cable
  OF paper-insulated lead-sheathed chloroprene-corrosion-protected steel-belt-armored cable
  OF paper-insulated aluminum-sheathed vinyl-corrosion-protected cable
    Conductor — compacted conductor, segmented compacted conductor, flat-type OF cable available

Pipe-type cable
  Pipe-type gas-compressed paper-insulated cable
  Pipe-type gas-filled paper-insulated cable
  Pipe-type high-pressure oil paper-insulated cable (POF cable)
    Conductor — compacted conductor, segmented compacted conductor

Polyethylene power cable
  XLPE-insulated vinyl-sheathed cable
  XLPE-insulated vinyl-sheathed steel-belt-armored cable
  XLPE-insulated polyethylene-sheathed cable
  PE-insulated vinyl-sheathed cable
  PE-insulated vinyl-sheathed steel-belt-armored cable
  Vertical riser PE-insulated vinyl-sheathed double-iron-wire-armored cable
  PE-insulated steel-belt-armored cable
  Under-water use PE-insulated single-iron-wire-armored cable
  PE-insulated lead-sheathed cable
  PE-insulated lead-sheathed steel-belt-armored cable
    Conductor — compacted conductor, segmented compacted conductor aluminum-sheathed type, corrosion-protected type, and corrugated-steel-tape-armored-type available

Butyl rubber power cable
  Butyl-rubber-insulated chloroprene-sheathed cable
  Butyl-rubber-insulated chloroprene-sheathed steel-belt-armored cable
  Shaft-use butyl-rubber-insulated chloroprene-sheathed double-iron-wire-armored cable
  Butyl-rubber-insulated vinyl-sheathed cable
  Butyl-rubber-insulated vinyl-sheathed steel-belt-armored cable

Table 2.1.1. (continued)
CLASSIFICATION OF POWER CABLES AND THEIR EXAMPLES

Butyl-rubber-insulated steel-belt-armored vinyl-corrosion-protected cable
Under-water use butyl-rubber-insulated single-iron-wire-armored cable
Butyl-rubber-insulated lead-sheathed steel-belt-armored cable
  Conductor — compacted conductor, segmented-compacted conductor

Ethylene-propylene rubber power cable
  EP rubber-insulated lead-sheathed cable, etc.
Forced-cooled OF cable

Forced-cooled POF cable (high pressure)

CGI cable (compressed gas insulated cable)
  Compressed-SF_6-insulated cable

Cryogenic resistive cable (CR cable)
  Liquid-nitrogen-impregnated paper-insulated cable
  Liquid-nitrogen-impregnated polymer-insulated cable
  Liquid-nitrogen-cooled vacuum-insulated cable

Superconducting cable (SC cable)
  Liquid-helium-insulated cable
  Liquid-helium-impregnated polymer-insulated cable
  Super-critical-helium-impregnated polymer-insulated cable

longitudinally folded and corrugated copper tape, applied with an overlap, under an extruded polyethylene overall jacket. This construction provides an excellent low impedance path for short circuit currents, equivalent to that of the lead sheath used on oil-impregnated paper cable or to that provided by skid wires and pipe in POF cable.

B. Design Stress and Insulation Thickness

   The choice of the design stress of a cable can be determined by either the impulse or the AC voltage requirement of the system in which the cable is installed. Two kinds of cables, taped cable and extruded cable, provide their respective methods for determining their insulation thickness. The value is usually determined from the basic insulation level, or BIL, which is meant to represent the maximum transient voltage that can occur, together with a suitable safety factor. BIL values are listed in Table 2.1.2.[2,3,9,10]

   A value of 1.2 (nondimensional) is recommended for the safety factor; this allows for uncertainties and the temperature factor. It is postulated in design of oil-filled taped cable that no temporal degradation takes place. It will be seen that in terms of working voltage, the impulse requirement becomes relatively less severe as operating voltage increases. It is, therefore, possible to increase the design stress of cable as the operating voltage increases.

   Insulation thickness of a taped cable is determined by the formula:[3]

$$t' = r \{\exp (V/rE) - 1\} \tag{2.1.2.}$$

where r = the radius of the conductor including its shielding, V = the impulse voltage level specified for the cable, and E = the maximum electric stress — the lowest value for impulse breakdown so far obtained. Nominal insulation thickness is the sum of the real insulation thickness and the conductor shield thickness. The calculated values are usually rounded up for safety reasons, and the resulting values are specified.

   As the impulse voltage is fixed for each operating voltage, all cables designed for that voltage have approximately the same stress. Reference to the formula for the stress in the insulation will show that as the conductor radius is increased the insulation thickness can be reduced while still maintaining the same design stress.

Table 2.1.2.
BIL VALUES IN THREE COUNTRIES

U.S.				Japan		U.K.	
Reference class (kV)	Standard BIL (kV)	Reduced BIL (kV)	Cable BIL (kV)	Reference class (kV)	BIL (kV)	Reference class (kV)	Cable BIL (kV)
1.2	30[a] 45[b]	—					
2.5	45[a] 60[b]	—		3.3	45		
5.0	60[a] 75[b]	—	60	6.6	60		
8.0			95				
8.7	75[a] 95[b]	—					
15	95[a] 110[b]	—	110	11	90		
23	150	—		22	150		
25			150				
28							
34.5	200	—		33	200	33	194
35			200				
46	250	—					
69	350	—		66	350	66	342
				77	400		
92	450	—		110	550		
115	550	450					
138	650	550	650[c]	154	750	132	640
161	750	650		187	750		
191	900	—		220	900		
230	1050	900		275	1050		
287	1300	—				275	1050
345	1550	—					
500			1550[d]	500	1550	400	1425
				1000	2250		

Note: U.S. Standard and reduced BILS: Chap. 18 (AIEE·EEI·NEMA, 1941);[2] U.S. Cable BILs (5 kV ~ 35 kV): AEIC No. 5-71; Japan BILs: JEC-169, Cable BIL = BIL × 1.2.

[a] For distribution class equipment.
[b] For power class equipment.
[c] ERC requirement for 138 kV XLPE cable at Waltz Mill.[9]
[d] EPRI requirement for 500 kV POF cable at Waltz Mill.[10]

For extruded cables, there are two formulas,[3] available to determine insulation thickness. One is based on impulse breakdown voltage and the other on AC breakdown voltage, as follows:

$$t_1 = \frac{\text{BIL} \times \alpha \times \beta}{E_1} + t_0 \qquad (2.1.3.)$$

where t_1 = the insulation thickness to be determined from the impulse breakdown voltage, BIL = the basic impulse level, α = the degradation factor, β = the temperature factor, t_0 = the thickness of the conductor shield, and E_1 = the mean electric stress for impulse breakdown voltage, and

$$t_2 = \frac{(E/\sqrt{3}) \times \gamma}{E_2} + t_0 \qquad (2.1.4.)$$

Table 2.1.3.
INSULATION THICKNESS OF
POWER CABLES[5-7,9-11]

System voltage (kV)	Insulation thickness (mil)	
	OF and POF cable	XLPE cable
15	100/115	175/220
25	135	260/300
35	170	345/420
69	285	—
138	505	800/835
230	760	—
345	1035	—
500	1340	—

Note: 100 mils = 25.4 mm.

where t_2 = the insulation thickness to be determined from the AC breakdown voltage, E = the system voltage, γ = the degradation factor, t_0 = the thickness of the conductor shield, and E_2 = the mean electric stress for AC breakdown voltage. In contrast with the taped cable described earlier, the extruded cable will degrade if any voltage is applied. The impulse voltage degradation factor α is therefore chosen by taking into consideration a number of effects. These are the impulse voltage and its repetition on impulse breakdown voltage, switching surge voltage, and sustained AC over-voltage. The AC voltage degradation factor γ may be determined from the V-t characteristic — a life-curve of the specified cable, or from data on the specified material when tested under a similar condition to actual service. Impulse breakdown voltage is temperature-dependent for polymer dielectrics. The temperature correction factor is applied since impulse breakdown tests are usually carried out at room temperature while the cable may be subjected to higher temperatures in actual service. For XLPE cables, typical values of α, β, and γ are 1.2, 1.3, and 4.0, respectively. These values may change through improvement of cable manufacturing processes, field experiences of the cables, and so on. If two values are obtained for the impulse and AC voltage requirements, the higher value is chosen. This is generally the impulse requirement. Examples of the insulation thickness available at present are shown in Table 2.1.3.[5-7,9-11]

C. Dielectric Properties

The capacitance of a cable per unit length is given by:

$$C = \frac{\epsilon}{18 \ln (R/r)} \quad (\mu F/km) \qquad (2.1.5.)$$

where ϵ = the specific dielectric constant of cable insulation, r = the conductor radius including the conductor shield, and R = the cable radius up to the outer surface of the insulation.

Electrical resistance of cable insulation is given by:

$$R_i = \frac{\rho}{2\pi} \ln (R/r) \times 10^{-11} \quad (M\Omega - km) \qquad (2.1.6.)$$

where ρ = the volume resistivity of the insulation (Ω-cm). It is to be noted that the product of the capacitance and the resistance of a cable is a parameter independent of cable length and radius. A value of 7200 ΩF is typical for OF cables. This parameter may be convenient to use as a degradation measure for insulation resistance because it is independent of cable dimensions and the capacitance is far less sensitive to degradation than the resistance.

Dielectric properties are a determining factor in cable design insofar as they dictate the charging current and the losses. Once a material has been selected, one is in large measure locked in. Losses determine the power rating of a cable; that is, its transmission capability. This is particularly true for taped cable and especially so for oil-filled type, Kraft® paper-insulated cable. As the operating voltage increases, so does the dielectric power loss, as indicated by the following formula:

$$\text{Dielectric loss} = V^2 \omega C \tan\delta \qquad\qquad (2.1.7.)$$

where V = the operating voltage, C = the electrostatic capacitance, ω = 2π times the frequency, and δ = the dielectric loss angle. It will be seen that the losses are proportional to the square of the operating voltage. However, this is partly mitigated by the fact that the insulation thickness increases as the operating voltage increases and this reduces the cable capacitance.

The transmission capacity is limited both by the charging current and by the thermal instability due to dielectric loss. The loss angle for oil/paper dielectric varies with temperature passing through a minimum at about 50 to 60°C. Thus, around the operating temperature, a rise in temperature increases the dielectric loss, which increases the heat generation. The rise in temperature will also increase the temperature gradient to the ambient and raise the rate of heat dissipation. If the rate of rise of heat dissipation does not keep pace with the increasing loss, the temperature will continue to increase until the insulation overheats and fails electrically. Typical values of dielectric properties are shown for various cables in Table 2.1.4.[3,12-16]

2.1.3. Thermal Design

A. Maximum Temperature

As in the case of most other electrical equipment, there is a maximum temperature at which the cable can be operated. This is chosen to be 80 to 95°C for impregnated, paper-insulated cables, with one exception, as shown in Table 2.1.5. Values of 75°C and 90°C are specified for PE and XLPE cables, respectively, as shown in Table 2.1.6.[7] These are defined as the hottest-spot temperatures that may be reached at any time during regular daily load cycle of a line. The current rating of the cable is then the maximum current that can be passed through the conductor without this temperature being exceeded. Limits for emergency operating temperatures apply only to the infrequent higher loading of a line necessitated by the unintentional loss of other lines or equipment. It is assumed that the shape of the daily load curve will be the same or similar to that under normal operation. The emergency temperatures are based on the assumption that their application will not exceed the requirements as shown in Table 2.1.7.[8] The third temperature is the maximum conductor temperature during a short circuit. These three maximum temperatures may be used in ampacity calculations provided there is adequate information about the overall thermal characteristics of the cable environment to assure that these temperatures will not be exceeded. In the absence of this information, the permissible temperatures should be reduced by 10°C, or in accordance with available data.

B. Heat Flow

The transmission of heat through the cable and its surroundings is by conduction in the case of buried cables, and the temperature difference across an element under steady state condition may be written as follows:[4]

$$\theta = W \cdot G \qquad\qquad (2.1.8.)$$

Table 2.1.4.
DIELECTRIC PROPERTIES OF VARIOUS LOW LOSS DIELECTRICS

	Specific weight (g/cm³)	ϵ_s	tanδ (%) 60 Hz	
HD polyethylene	0.92	2.2 —2.3	0.03	
M & LD polyethylene	0.95	2.2 —2.4	0.03	
PE cross-linkable compound	0.93—1.40	2.28—7.60	0.3—4.4	
Polypropylene	0.90	2.0 —2.2	< 0.05	
Polystyrene	1.0	2.3 —2.7	0.01—0.03	
Polytetrafluoroethylene	2.2	2.0 —2.1	0.09	
Polycarbonate	1.2	2.9 —3.1	0.05—0.09	
Polysulfone	1.1	3.1	0.08	
EEA copolymer	0.93	2.7 —2.9	1—2	
EVA copolymer	0.935—0.950	2.5 —3.16	0.3—2.0	
Polypropylene paper	0.75	2.2	0.03	Polybutene-impregnated
Polyethylene paper	0.37	1.9	0.03	Polybutene-impregnated
Polyester paper	0.68	2.1	0.45	Polybutene-impregnated
Polycarbonate paper	0.75	2.65	0.045	DDB impregnated
P₃O paper	0.65	2.5	0.025	Fluid paraffin-impregnated
		Viscosity (cst 80°C)	Gas absorption	
Mineral oil	0.89	3.4	− 35	
DDB	0.87	2.5	− 190	(+):Gas exhalation
Low viscosity polybutene	0.84	6.8	− 180	(−):Gas absorption
Fluid paraffin (I)	0.88	5.8	+ 22	
Fluid paraffin (II)	0.88	3.3	− 18	

where θ = temperature difference across the element (°C), W = heat flow through the element (watts), and G = thermal resistance of the element (°C/W). There are three principal sources of heat dissipation in a cable: (1) conductor losses W_c, (2) dielectric losses W_D, and (3) sheath losses W_s. The conductor temperature θ_c may then be written:

$$\theta_c - \theta_A = \left(W_c + \frac{W_D}{2}\right) G_1 + (W_c + W_D + W_s) G_2 \qquad (2.1.9.)$$

where θ_A = ambient temperature (°C), G_1 = internal thermal resistance of the cable (°C/W), and G_2 = thermal resistance of the ground (°C/W).

The above formula assumes that the dielectric losses are generated in the middle of the cable insulation. This equation must be solved to determine the current rating of the cable. With the exception of G_2, the various factors can be determined quite accurately. The thermal resistance of the ground requires the use of formulas involving estimated or experimentally determined values of the thermal resistivity of the undisturbed ground. To some extent this value can be estimated by considering the nature and structure of the ground, but in practice each value is subject to major variation by such influences as rainfall, drainage, height of water level, and type of ground surface. For a pipe-type cable, the pipe loss should be added to the second term of the Equation (2.1.9.). Typical values of thermal resistivity are tabulated for various materials in Table 2.1.8.[12,13,17-19]

C. Proximity Effect

When one or more sets of three single-core cables are laid together, the values of the factors in the heat equation are considerably modified. The magnetic fields of the conductor currents interact and disturb the even distribution of current across each conductor cross section. This proximity effect causes an increase in the AC resistance of the conductor, the increase being smaller as the cable spacing is increased. At first sight, it might appear that

Table 2.1.5.
AEIC MAXIMUM CONDUCTOR TEMPERATURES FOR IMPREGNATED-PAPER-INSULATED CABLE — CABLE MANUFACTURED DURING 1967 AND LATER

	Conductor Temperature,° C	
Rated voltage (kV)	**Normal operation**	**Emergency operation**

Solid-Type Multiple Conductor Belted

1	95	115
2 to 9	90	110
10 to 15	80	100

Solid-Type Multiple Conductor Shielded and Single-Conductor

1 to 9	95	115
10 to 29	90	110
30 to 49	80	100
50 to 69	65	80

Low-Pressure Gas-Filled

10 to 29	90	110
30 to 46	80	100

Low-Pressure Oil-Filled and High-Pressure Pipe Type

		100 Hour[a]	**300 Hour[a]**
15 to 39	95	115	110
40 to 345	85	105	100

Note: These tabulated temperatures are recommended as permissible conductor temperatures for ampacity calculations when adequate knowledge of the overall thermal characteristics of the cable environment is available. Lower temperatures should be used when adequate knowledge of thermal conditions is not available.

[a]   Use temperature limits which correspond to the expected maximum duration of any emergency operation.

Table 2.1.6.
MAXIMUM CONDUCTOR TEMPERATURES FOR PE AND XLPE CABLES RATED 5 TO 69 kV

Type insulation	Normal operation	Emergency operation	Short circuit operation
Polyethylene	75	90	150
Cross-linked polyethylene	90	130	250

Table 2.1.7.
REQUIREMENTS FOR EMERGENCY
TEMPERATURES

| Type of cable | Maximum length of each period (hr) | Maximum number of emergency periods | |
		In any 12 consecutive months	Average per year for life of cable
Solid	36	3	1.0
Low-pressure gas-filled	36	3	1.0
Low-pressure oil-filled	100 or 300	1	0.2
High-pressure pipe	100 or 300	1	0.2

this proximity effect would be worse on three core cables where the conductors are in the closest proximity. While this is true up to a point, the proximity effect is also very dependent on the conductor size and, in the case of single core cables, the maximum conductor size is about five times larger than that used for three core cables. The magnetic fields also induce longitudinal voltages in the cable sheaths and circulating currents will flow if the sheaths are bonded together at the ends of the cable route. The magnitude of the circulating currents, and of the heat losses they produce, will increase as cable spacing is increased. Finally, the heat losses of the three cables share to some extent the same heat path through the ground, this mutual heating effect decreasing as cable spacing is increased. Thus, the heating equation factors are not only modified, but interact, so that for any specific duty, it is necessary to calculate a number of conductor size/cable spacing alternatives to determine the economic arrangement.

D. Metal Sheath[3,20]

Oil-impregnated paper-insulated cable usually has a lead or aluminum sheath, as stated before, to provide a hermetic seal against the absorption of moisture. Consideration of mechanical requirements sets a minimum limit to the thickness of the lead sheath. This in turn leads to values of electrical and thermal capacitance which give the lead sheath a quite high short circuit rating. As a consequence, the short circuit rating of the sheath is not a limiting factor, except in the case of cables with very large conductors.

By contrast, most extruded cables are not provided with a hermetic seal in the belief that the insulation will be able to operate in the presence of moisture and air although it is evident that water will affect the insulation characteristics, resulting in the formation of water trees or electrochemical trees in extreme cases. Several forms of metal sheaths are available as follows:

1. One or two metal tapes applied helically with a lay sufficiently short to provide a measure of flexibility
2. Helically applied tape shields with two tapes applied with opposite lay
3. Helically applied copper wires covered with semiconducting plastic
4. Cylindrical metal sheaths

The universally accepted method used for estimating the short circuit rating of electrical conductors is to assume that the heat generated by the short circuit operates upon the thermal capacitance of the conductor to raise its temperature without dissipation of heat to its

Table 2.1.8.
SPECIFIC THERMAL RESISTIVITY

Material	Specific thermal resistivity (°C-cm/W)	Material	Specific thermal diffusion resistivity (°C cm/W)
Oil-impregnated paper		Lead sheath	$1300 \ (500 + 20d_s, \ d_s \leq 40)$
		Jute	$900 \ (500 + 10d_s, \ d_s \leq 40)$
Solid	700	Chloroprene	$900 \ (500 + 10d_s, \ d_s \leq 40)$
Low gas pressure	700	Vinyl	$900 \ (500 + 10d_s, \ d_s \leq 40)$
OF	550	Polyethylene	$900 \ (500 + 10d_s, \ d_s \leq 40)$
PGF	700	Impregnated braid	$800 \ (400 + 20d_s, \ d_s \leq 20)$
POF, PGC	550	Pipe cable	
		With corrosion protecting layer	1000
		Without corrosion protecting layer	1300
Natural rubber	500	Trough	1000
Butyl rubber	500		
EP rubber	500		
Chloroprene rubber	500		
Polyethylene	450, 350, 400		
XL polyethylene	450, 350		
Vinyl	600		
Cambric	600		
Jute	600		
Pipe corrosion protecting layer	600		
Porcelain insulator	600		
Asbestos	600		
Average over soil and pipe			
Wet soil	60		
Common soil	100		
Dry soil	150—300		
Sand and trough	200		
Concrete			
Light	110		
Heavy	70		
Water			
30°C	165		
70°C	150		

Note: d_s = Outer diameter of cable or pipe. Values for concrete, sand, and trough may change according to the content of water.

environment. The initial temperature of the conductor is assumed to be that established by the normal full load rating of the cable and the final temperature is specified so as to avoid damage to the insulation or protective coverings with which the conductor is in contact. In the U.S., the same methods are applied to metallic shields in the Insulated Power Cable Eng. Assoc. (IPCEA) publication relating to this subject.[27]

Noting the deviation of values calculated in this way from experimental values, a new formula was proposed for helically applied copper tape in 1968, and since then has been widely used. This was developed on the basis of five simplifying assumptions as follows:

1. The conductor is plane rather than cylindrical
2. The temperature of the conductor is uniform
3. The adjacent thermal insulator, represented by the cable insulation and jacket extends to infinity
4. The adjacent cable components are in good thermal contact with the metal shield
5. The thermal coefficients of the materials remain constant

It is still difficult in this treatment to choose the most suitable electrical resistance of copper tape and the permissible temperature. Actual values deviate from the resistance obtained for a thin copper cylinder, and, therefore, a geometrical coefficient should be considered. The permissible temperature may be the short circuit permissible temperature of the insulation, the short circuit permissible temperature of plastic sheath, or a temperature in between.

Fault current capacity for extruded cables can be obtained from the following formula,[20]

$$T = \frac{a}{(R_1 - R_2)M} \left[\frac{1}{R_1} \left(e^{R_1^2 t} \, \text{erfc} R_2 \sqrt{t-1} \right) - \frac{1}{R_2} \left(e^{R_2^2 t} \, \text{erfc} R_2 \sqrt{t-1} \right) \right] \qquad (2.1.10.)$$

where T = the maximum temperature rise at flat time (°C), a = watts generated in the shield when $t = 0$ per unit area of surface which dissipates heat (W/cm^2), b/a = temperature coefficient of shield resistance, M = thermal capacitance of shield per unit area of surface which dissipates heat (J/°C·cm^2), k = thermal conductivity of the medium adjacent to the shield (W/°C·cm), K = diffusivity of the medium $-kc\rho$(cm^2/s), c = thermal capacitance of the medium adjacent to the shield (J/°C·g), and ρ = density of the medium adjacent to the shield (g/cm^3), and

$$R_1 = \frac{k}{2M\sqrt{K}} + \sqrt{\frac{k^2}{4M^2K} + \frac{b}{M}} \qquad (2.1.11.)$$

$$R_2 = \frac{k}{2M\sqrt{K}} - \sqrt{\frac{k^2}{4M^2K} + \frac{b}{M}} \qquad (2.1.12.)$$

The formula does not give directly the short circuit current rating of a shield corresponding to a given temperature rise. However, by judicious selection of two or more values of short circuit current, one can find the relationship between temperature rise and short circuit current of the selected duration and interpolate to find the rating.

E. Cross Bonding of Sheaths

The current rating of the single-core cable system can be substantially increased if the losses due to sheath circulating currents can be eliminated. Except for short cable routes, this is achieved by cross bonding the sheaths of the cables at the joint positions so that over a section consisting of three drum lengths of cable, the overall induced voltage is approximately zero.

F. Forced-cooled System

Forced-cooled cable systems will provide a substantial increase in transmission power. This is and will be used for extruded cables as well as POF and OF cables. Heat generated in the conductor by Joule heating and in insulation by the dielectric heating is carried away by flowing oil or water internally or externally. This technique will also tend to smooth out hot spots generated by thermal inhomogeneities along the cable length. Liquid nitrogen impregnated cable — a new cable — is a forced-cooled cable in a sense.

2.1.4. Mechanical Design

As one of the fundamental requirements for a cable is an ability to bend during manufacture and laying without damaging the insulation, it would be ideal if the diameter of the reel or capstan on which the cable is wound were large enough to ensure that the deformation of the insulation during the bending could be maintained within the elastic limit of an insulating material selected.

A. Bending Performance

With this in mind, one might attempt for OF cable to obtain such compact lapping that there would be no tape movement during bending, the friction stresses exceeding the elastic limit. Unfortunately, if this route is followed the necessary winding diameter would be quite impractical, and, therefore it has to be accepted that the papers must slide with respect to one another during bending. Owing to the radial pressure caused by the lapping operation and bending, friction forces are created which, on bending the cable, stress the paper tapes in the slipping direction. In the compressed part of the cable, these forces exert an end thrust on each single paper turn, and in this manner they can set up critical conditions which generally cause some damage to the insulation; the damage can be described as wrinkles or creases according to whether the edges are rounded or sharp.[21]

Some types of damage to the insulation as a result of bending are represented in Figure 2.1.2. (a, b, and c).[21] Figure 2.1.2. (a) shows the butt-space creases which occur in the compressed part of the insulation where the friction forces necessary for the sliding of the paper tapes are such as to cause the collapse of the paper as end thrust corresponds with the butt spaces. Collapse creases as shown in Figure 2.1.2. (b) occur in the compressed zone when the paper tapes also collapse outside the butt spaces. They are characterized by the total collapse of a considerable thickness of insulation and appear in the form of deep buckling with rather sharp edges. Collapse wrinkles as shown in Figure 2.1.2. (c) occur in the compressed zone of the insulation owing to defective compactness. They are the result of the collapse of single paper turns due to the axial and radial forces acting on the paper tape during bending.

It has been demonstrated experimentally that the formation of creases and/or wrinkles reduces breakdown strength. The cable insulation must therefore be designed in such a way as to avoid the formation of creases and wrinkles. Radial pressure should be limited to a critical upper pressure in order to prevent any creases, while it should be increased up to a critical lower pressure in order to suppress butt spaces between adjacent tapes. Lapping condition may be determined from the upper and lower pressures and of course from past experience.

To prevent cable insulation from wrinkling, it is necessary to increase the Young's modulas of the cable and decrease the friction at the paper surface. As Young's modulas is comparatively low for EHV cable which uses low density paper, an attempt has been made to develop mica-loaded paper[22] to mitigate this shortcoming. It is generally required that a single-core cable withstand a bending radius equivalent to 20 to 25 times the cable radius; the corresponding figures for a 3-core cable are 12 to 15 times the cable radius.

POF cable is constructed with a metallic insulation shield over which are placed successively a moisture-proofing layer, a reinforcing layer, and skid wires, but no metallic sheath is used. The lapping condition for this type of cable is determined in a similar way to that for OF cable.

Extruded cable can, by its nature, withstand more bending than taped cable, but extreme bending, say less than six times cable-radius bending, will cause its electrical performance to deteriorate. A permissible bending radius greater than ten times cable radius is usually chosen.

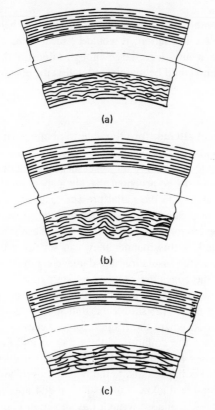

(a) Butt-space creases (paper tapes staggered by 30%)
(b) Collapse creases (tapes staggered by 50%)
(c) Collapse wrinkles (tapes staggered by 50%)

FIGURE 2.1.2.   Formation of creases and wrinkles in OF cable insulation. (From Gazzana Priaroggia, P., Occhini, E., and Palmieri, N., IEEE Monograph No. 390 S, July, 2, 1960, p.2. With permission.)

B. Lateral Pressure Performance

Cable subjected to excessive lateral pressure at the end of pull-in is liable to damage. Aluminum-sheathed OF cable is three times stronger in compression than lead-sheathed OF cable. The permissible lateral pressure in cable installation may be determined by considering the inner-surface conditions of the pipe and the abrasive property of the protective layer. A value of 400 lb/ft or 600 kg/m has been adopted as the permissible lateral force for POF cable in the U.S.

Corrugated steel pipe is stronger in compression than sheaths of aluminum, lead, or lead sheath with steel belt. It is also light and easy to handle. For these reasons it has been utilized recently for low-voltage, direct-buried extruded cable.

Electromagnetic forces due to short circuits and impact forces experienced in service should be taken into consideration in cable design. Metal sheaths of OF cable and steel pipe of POF cable should be designed to meet these and other mechanical requirements to be discussed in the next section.

2.1.5. Hydraulic Design[4]

A. OF Cable

OF cables require an oil feeding system to compensate for variations in oil quantity and pressure as the temperature varies. At no point in the insulation of an oil-filled cable should

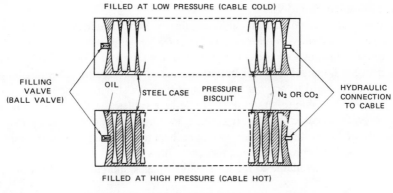

FILLED AT LOW PRESSURE (CABLE COLD)

FILLING VALVE (BALL VALVE)   OIL   STEEL CASE   PRESSURE BISCUIT   N₂ OR CO₂   HYDRAULIC CONNECTION TO CABLE

FILLED AT HIGH PRESSURE (CABLE HOT)

OIL CAPACITY DETERMINED BY RESERVOIR LENGTH.

FIGURE 2.1.3.   Sketch of cross-section of oil-filled cable oil reservoir (From Arkell, C. A., *Underground High Voltage Power Cables* BICC Cable Division Post Graduate Education Centre, United Kingdom Atomic Energy Assoc., Harwell, Buckinghamshire, England, 1968. With permission.)

the oil pressure ever fall below atmospheric pressure. A further essential requirement is that the insulation contain the minimum practicable quantities of air and moisture.

An important component is the pressure tank which consists of a number of circular "biscuits" enclosed in a cylindrical container; the general arrangement is shown in Figure 2.1.3.[4] Each biscuit consists of two circular diaphragms connected together at their edges by a short cylinder. Before insertion into the tank, the biscuits are filled with gas, at 1 atm for standard tanks and at 1.5 or 2 atm for prepressurized tanks. Following mechanical assembly, the spaces between biscuits and tank is vacuum-filled with dry, degassified cable oil. Raising the external oil pressure causes the two diaphragms of each biscuit to flex towards each other, reducing the volume of the biscuit and compressing the gas within it. Oil enters the container to occupy the contraction volume of the biscuits. If the external oil pressure is lowered, the compressed gas within the biscuits expands them, driving oil out of the tanks.

B. POF Cable

POF cable is operated at pressure as high as 15 kg/cm² or 200 psi. in its pipe. This needs a considerable amount of oil which consequently requires an abundance of compensating oil.[3] It is uneconomical to utilize the oil tank used for this purpose with OF cables. A pumping plant, which includes a pressurizing pump, normally operates to follow oil expansion and contraction. An oil storage tank contains sufficient oil to compensate for expansion and contraction of the oil in the cable system, plus an extra quantity for emergency. Oil is supplied by an oil hydraulic pump which partially returns it to the storage tank through a relief valve when temperature rise causes an increase in pressure. In this way the oil pressure inside the pipe may be maintained within specified limits.

Longitudinal oil-flow resistance is lower in POF cable than OF cable. Therefore, in many cases, one pumping plant is sufficient to supply oil to one cable system, and even multiple circuit cables in parallel, or radial and looped installations. Considerations of reliability may dictate an additional redundant pumping plant.

OF cable and POF cable are manufactured in discrete lengths and then coiled onto drums. Such accessories as straight joints, trifurcating joints for three-core OF cable, and terminations are required to complete a cable installation.

In addition, due to the use of a pressurized fluid impregnant, several special purpose accessories are required, notably the stop joints which contain hydraulic barriers.

2.1.6. Cross-Bonded Sheaths

Electric potentials are induced on cable sheaths by electromagnetic induction. The permissible voltage is normally in the range of 30 to 60 V, being determined by such factors as the dielectric breakdown of corrosion-protective coverings and the safety of personnel. It may be between 60 and 100 V for buried sheaths, or for sheaths at terminations which are provided with a guard to prevent hand contact with the sheath or metal parts connected thereto. It is most desirable to strand three single-core cables. As this is difficult or nearly impossible in practice, cable sheaths are earthed, leading to current flow along them. These sheath currents are undesirable because they cause heat losses which reduce the cable rating; they may also interfere with communication cable lines.

Several methods are available to reduce sheath voltages and currents. They are

1. Transposition
2. Single point bonding
3. Solid bonding
4. Cross bonding — sectionalized and continuous
5. Other methods such as low-resistance earthing, reactance bonding, or transformer bonding

The voltage gradient induced in a cable sheath is greatly affected by the cable formation, that is to say whether it lies in a trefoil or flat formation, and the spacing between paralleled cables. It is at a minimum for the compact trefoil formation.

Any conductor P, lying parallel with a set of three conductors carrying balanced three-phase current, will have a voltage gradient induced along its length, given by

$$E_p = j\omega I \times 2 \times 10^{-7} \left(\frac{1}{2} \ell n \frac{S_{1p} S_{3p}}{S_{2p}^2} + j \frac{\sqrt{3}}{2} \ell n \frac{S_{3p}}{S_{1p}} \right) \quad (\text{v/m})$$

(2.1.13.)

where I = the rms current in amperes in conductor no. 2, ω = the angular frequency of the system, S_{1p} = the axial spacing of the parallel conductor and phase 1 conductor, S_{2p} = the axial spacing of the parallel conductor and phase 2 conductor, S_{3p} = the axial spacing of the parallel conductor and phase 3 conductor. Voltage gradients can be obtained for the trefoil formation case by putting $S_{12} = S_{23} = S_{13} = S$ and $S_{11} = S_{22} = S_{33} = d$, and for flat formation case by putting $S_{12} = S_{23} = S$, $S_{13} = 2S$ and $S_{11} = S_{22} = S_{33} = d$.

Voltage induction in parallel cables resulting from balanced loads can be reduced or eliminated by transposition. For this reason transformation is sometimes employed together with one of the specially bonded cables.

Short cable circuits, such as interconnections within a substation or terminations of an overhead line into a substation are frequently handled by bonding and earthing the three sheaths at only one point along the cable route. This point is normally at one end or at the center. It is clear that a voltage will appear from sheath to earth which will be a maximum at the farthest point from the earth bond. The sheaths must therefore be adequately insulated from earth. Since there is no closed sheath current except through the sheath voltage limiter (if any), current does not normally flow longitudinally along the sheaths and no sheath circuit loss therefore occurs. Sheath eddy current loss will, of course, still be present. Since this sheath cannot carry any of the returning current, yet is affected by the occurrence of an earth fault in the immediate vicinity of the cable, it is recommended that a single point bonded cable installation be provided with a parallel continuous conductor which is earthed at both ends of the route.

When sheaths are continuous and earthed at both ends of the route, they act as screening conductors and thus reduce somewhat voltages induced in parallel cables.

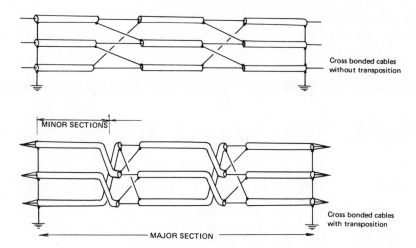

FIGURE 2.1.4.   Cross bonded cable systems. (From CIGRE SC-21 WGO 7-38 Document 07-38, *The Design of Specially Bonded Cable Circuits (Part II)*, CIGRE, Paris, 1974, p. 65. With permission.)

Cross bonding is now widely employed for longer cable circuits. It is essentially constructed by sectionalizing the sheaths into elementary sections and cross connecting them so as to approximately neutralize the total induced voltage in three consecutive sections. With nontransposed cables, it is impossible to achieve an exact balance of induced sheath voltages unless the cables are laid in trefoil. It is therefore desirable to transpose the cables at each joint position. When this is done the induced sheath voltages will be neutralized, irrespective of cable formation, provided the three elementary sections are identical. Figure 2.1.4.[28] shows how this can be done for a circuit consisting of three elementary sections only. The sheaths are bonded and earthed at both ends of the route. The phasor sum of sheath voltages in this unit of three minor sections is zero, thus sheath losses are reduced. Maximum sheath voltage can be determined by considering one unit.

There are several ways available, as described below[28] to connect one unit to another so as to construct longer cable circuits.

A. Sectionalized Cross Bonding

If the number of elementary sections is exactly divisible by three, the circuit can be arranged with one or more major sections in series. At the junction of two major sections, and at the ends of the circuit, the sheaths are bonded together and earthed, but generally the earths at the junctions of major sections will be only local earth rods. In Figure 2.1.5.[28] each separate major section is connected as in Figure 2.1.4. Sheath voltage limiters, if required, are added at the cross-bonded joints only.

B. Continuous Cross Bonding

In this system the sheaths are cross-bonded at the end of each elementary section throughout the entire cable route. The three sheaths are bonded and earthed at the two ends of the route only, as shown in Figure 2.1.6.[28] Again, it is generally desirable that cables be transposed so that each conductor occupies each of the three positions for one third of the total length. It is preferable that the number of matched elementary sections be exactly divisible by three, but if some unbalance can be tolerated, any number may be used. Clearly, the unbalance created by using a number of elementary sections not divisible by three decreases as the total number of sections increases.

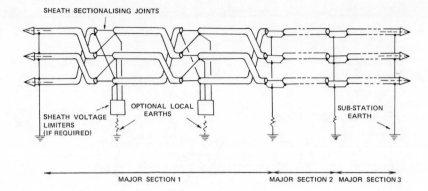

FIGURE 2.1.5. Cross bonded cables (sectionalized bonding) with three major sections. (From CIGRE SC-21 WGO 7-38 Document 07-38, *The Design of Specially Bonded Cable Circuits (Part II)*, CIGRE, Paris, 1974, p. 66. With permission.)

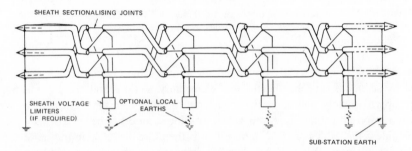

FIGURE 2.1.6. Continuous cross bonding. (From CIGRE SC-21 WGO 7-38 Document 07-38, The Design of Specially Bonded Cable Circuits (Part II), CIGRE, Paris, 1974, p.66. With permission.)

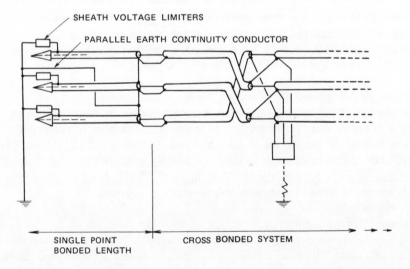

FIGURE 2.1.7. Termination of cross bonded system with single point bonded length. (From Mildner, R. C., *IEEE Trans. Power Appar. Syst.*, 87(3), 749, 1968. With permission.)

C. Mixed Systems

When the number of elementary sections is not exactly divisible by three, the system may be designed to contain a mixture of cross-bonded and single-bonded lengths. Figure 2.1.7.[28] shows the arrangement of a final single point bonded length at the end of a cross-bonded system of either the sectionalized or the continuous type.

In a cross-bonded system, the power cable sheaths do not carry currents when balanced, hence the screening effect is absent. When supplying unbalanced loads or during faults, sheath currents will flow and therefore an important screening effect is present.

There is generally some unbalance in practical systems due to the difficulty in constructing identical phase lengths and constant spacings. Wherever this occurs the sheath loss should be in the permissible range.[32]

D. Power Frequency Overvoltages

System faults produce an initial transient overvoltage, followed by a power frequency sheath overvoltage caused by the passage of the fault current. This power frequency overvoltage is not generally high enough to influence the sheath insulation design, but since it persists for the duration of the fault it may be important in relation to the duty requirements of sheath voltage limiters. This should be considered for three types of fault: three-phase symmetrical fault, phase-to-phase fault, and single-phase earth fault.[28]

E. Transient Overvoltages

A three-phase single conductor cable system with cross bonding may produce extraordinary transient voltages when subjected to lightning or switching surges, because a cross-bonding point acts as a singular point for traveling wave propagation. The following results were obtained from both theoretical and experimental investigation:[3,33]

1. When a conductor is subjected to a surge with a fast-rising wave front, a voltage corresponding to 15 to 20% of the surge appears between the sheath and ground, while a voltage twice as great as the surge can appear at both ends of insulated joint boxes.
2. The surge wave attenuates drastically towards the center of the line involved, and damps exponentially with distance due to the earthing effect of ordinary joint boxes.

Breakdown of protective coverings or insulated joint boxes is probable for 138 kV or 154 kV class of cables.

It is in principle possible to calculate transient overvoltages in a specified system.[29] It is useful to summarize the results of measurements of sheath voltages in actual buried cross-bonded cables which have been subjected to incoming transient voltage waves; this is done in Table 2.1.9. It can be seen that for a variety of cables, the maximum voltage measured across the joint sleeve sectionalizing insulators varies from 25 to 61% of the incoming surge. The maximum voltage normally occurs at the cross-bonded joint nearest to the termination at which the surge enters the system.

Recommendations for insulation coordination of specially bonded systems protected by sheath voltage limiters are given in Table 2.1.10. They were derived by establishing appropriate values for various parameters: maximum protective level for sheath voltage limiters (1 if delta-connected or 2 in series if star-connected) plus a safety margin; i.e., 25 kV, Z_1 = 20 Ω, Z_2 = 10 Ω; a polyethylene-insulated bonding lead with a surge impedance of 30 Ω; and velocity of propagation of 300 m/μsec.

Sheath losses can be calculated with the aid of formulas developed by Miller and Imai.[35,36]

2.2. CABLE STRUCTURAL MATERIALS

2.2.1. Conductors

A. Required Performances

Annealed copper and hard-drawn aluminum are normally used for cable conductors. Their selection was based on the following considerations: (1) high electrical conductivity, (2) mechanical strength and flexibility, (3) ready workability and ease of handling, (4) chemical stability, and (5) economic issues such as availability, initial cost, operating cost, and salvage value. Sodium is now under consideration. Various performance characteristics of the above three metals are tabulated in Table 2.2.1.[3,12,13,23]

Table 2.1.9.

RESULTS OF MEASUREMENTS OF SHEATH OVERVOLTAGES IN BURIED CROSS-BONDED CABLE CIRCUITS WITH SHEATH VOLTAGE LIMITERS

Date	Source of information	Cable and installation details	Input voltage	Voltage across 1st sheath interruption (%)	Voltage sheath to earth (%)	Z_1	Z_2	Ratio Z_1/Z_2	Ref.
1959	Haga, Kusano, Japan	70 kV 400 mm² OF neoprene oversheath cables in ducts	1/1,700 μsec	46	−21 +24	14.3	9.9	1.45	33
1963	Watson, Erven, Canada	230 kV OF cable direct buried	Approx. step function	26	−14 +12	23.1	7.5	3.1	34
1964	Takahashi, Japan Report to Study Committee No. 21	154 kV 1200 mm² OF neoprene oversheath cables in ducts	3/9 μsec	25	−14.2 +12.7	15.3	4.7	3.25	
1965	Ball; Occhini, Luoni. Tests in U.K.	275 kV 1100 mm² OF direct buried at 130 mm centers, flat	Step function	40	−20 +20	19.0	10.9	1.75	30
1969	Provoost; Janssen, Holland. Report to S.C. 21	150 kV OF direct buried		50		23	18.2	1.26	
1969	Petry, Vierfuss, Germany. Report to S.C. 21	1 kV 1400 mm² OF: direct buried at 200 mm centers flat	Step function	61	−30.5 +30.5	10.0	11.5	0.87	
1971	Pluvinage; France. Report to working group	225 kV 410 mm² OF corrugated alum sheath, direct buried in trefoil	Impulse	31	−15 +15	30	12.1	2.48	
1972	Bazzi; Italy. Report to working group	220 kV 1200 and 1800 mm² direct buried at 350 mm flat	Step function	52	−28 +26	14.5	12.3	1.18	

Table 2.1.10.
RECOMMENDED INSULATION LEVELS FOR SPECIALLY BONDED SYSTEMS PROTECTED BY SHEATH VOLTAGE LIMITERS

| Highest voltage for equipment Um (rms) | Rated lightning impulse voltage (peak) | Minimum insulation level of joint sleeve[a] sectionalizing insulator 1.2/50 wave | |
		Bonding lead not exceeding 3 m	Bonding lead not exceeding 10 m
(kV)	(kV)	(kV)	(kV)
36	170	40	40
52	250		
72.5	325	40	40
123	450		
	550	40	60
145	650		
245	750		
300	1050	60	95
362	950		
420	1425	75	125
525	1175		
	1550	75	145

[a]   The insulation level of the inner insulation of the concentric bonding lead should be the same as for the joint sleeve sectionalizing insulator. The insulation level of the cable sheath and joint sleeves to ground and of the outer conductor of the bonding lead may be taken as half these values.

Table 2.2.1.
SOME PHYSICAL PARAMETERS FOR CABLE USE CONDUCTOR MATERIALS

	Annealed copper wire	Hard-drawn aluminum wire	Sodium wire
Chemical symbol	Cu	Al	Na
Atomic weight	63.57	26.97	22.79
Density (g/cm³)	8.89	2.70	0.972
Melting Point (°C)	1083	652—657	97.8
Specific heat (at 20°C)	0.092	0.225	0.295
Linear expansion temperature coefficient (1/°C at 20°C)	16.8×10^{-6}	23.6×10^{-6}	71×10^{-6}
Thermal conductivity (cal/cm·sec °C)	0.934	0.55	0.32
Tensile strength (kg/mm²)	24	16	
Elongation (%)	38.5	6	
Elasticity coefficient (kg/mm²)	11.9×10^3	6.3×10^3	
Specific resistivity (μΩ-cm)	1.71	2.83	4.2
Electrical conductivity (IACS %)	100	61	
Specific permittivity	0.9999905	1.0000208	

Considerations of availability and cost have led to increased use of aluminum where the larger conductor size is not a handicap. The electrical conductivity of aluminum is critically dependent upon minute chemical and metallurgical impurities difficult to eliminate in practical production. Impurities such as Ti and Mn will cause a tremendous increase in electrical resistivity of aluminum. Hard-drawn aluminum is preferable to annealed aluminum because of its higher tensile strength. Annealed aluminum has approximately 40% of the tensile strength of annealed copper.

Sodium is characterized by high electrical conductivity, low density, and low cost, which are all favorable for use in cable conductors.[38-52] Unfortunately, sodium is chemically unstable. It will easily oxidize in a normal atmosphere and will react violently with water. Tests have been made to use sodium in the form of an extruded polyethylene-insulated sodium-conductor cable.

B. AC Resistance of Stranded Conductors[1]

A conductor offers a greater resistance to the flow of alternating current than it does to direct current. The magnitude of the increase usually is expressed as an "AC/DC ratio". The reasons for the increase are several, the principal one being skin effect, the tendency for alternating current to crowd toward the surface of the conductor. Other contributing factors are proximity effect, the distortion of current distribution due to the magnetic effects of other nearby currents, and hysteresis and eddy current losses in nearby ferromagnetic materials and induced losses in short circuited nearby nonferromagnetic materials. Table 2.2.2. tabulates the AC/DC ratio including skin effect, proximity effect, and magnetic effect for some common sizes, according to IPCEA standards.[1,24]

It is well known that alternating current tends to flow more densely near the outer surface of a conductor; hence, the name "skin effect". The resulting reduction in effective conductor area, of course, increases the apparent resistance of the conductor. The flux linking a conductor (current) due to nearby current distorts the cross sectional current distribution in the conductor in the same way as the flux from the current in the conductor itself. The former is called proximity effect.

AC apparent resistance can be calculated by the formula

$$\gamma = \gamma_0 \times k_1 \times k_2 \qquad\qquad (2.2.1.)$$

where γ_0 = DC resistance (Ω-cm) at 20°C, k_1 = the ratio of DC resistance at the maximum permissible conductor temperature to that at 20°C, and k_2 = AC/DC resistance ratio.

$$k_1 = \{1 + \alpha (T_1 - 20)\} \qquad\qquad (2.2.2.)$$

where α = temperature coefficient of resistance, and T_1 = the maximum permissible conductor temperature.

$$k_2 = (1 + \lambda_s + \lambda_p) \qquad\qquad (2.2.3.)$$

for extruded and OF cables and

$$k_2 = \{1 + \beta (\lambda_s + \lambda_p)\} \qquad\qquad (2.2.4.)$$

for POF cable, where λ_s = skin effect coefficient, λ_p = proximity effect coefficient, and β = POF cable coefficient, β = 1.7 (regular triangle arrangement and β = 2.0 (cradle arrangement). Values of the temperature factor k_1 are given for copper and aluminum conductors in Table 2.2.3.[3]

C. Estimation Formula of the Skin and Proximity Effects

The skin effect coefficient λ_s and the proximity effect coefficient λ_p can be obtained from the following formulas,[3] which are dependent on the shape of conductors.

$$\lambda_s = F(X) = \frac{X(\text{ber } x \text{ bei}' x - \text{bei } x \text{ ber}' x)}{2\{(\text{ber}' x)^2 + (\text{bei}' x)^2\}} \qquad\qquad (2.2.5.)$$

Table 2.2.2.
DC RESISTANCE, AC/DC RESISTANCE RATIO OF CABLE CONDUCTORS

Spaced single conductor cable skin effect resistance ratio at 60 cycles and 65C copper
and aluminum conductors

(1)	(2)	(3)	(4)	(5)	(6)	(7)
	Concentric Round		Compact Segmental		Annular	
Cdr size AWG C.I.	DC Res 65 C	Skin effect resistance ratio	DC Res 65 C	Skin effect resistance ratio	DC Res 65 C	Skin effect resistance ratio
Copper Conductors						
4/0	58.9	1.01*	—	—	—	—
0.25	49.7	1.01*	—	—	—	—
0.30	41.5	1.01	—	—	—	—
0.35	35.5	1.01	—	—	—	—
0.40	31.2	1.01	—	—	—	
0.50	24.9	1.02	—	—	—	
0.60	20.8	1.03	—	—	—	—
0.70	17.8	1.03	—	—	—	
0.75	16.6	1.04	—	—	16.9	1.02
0.80	15.6	1.04	—	—	15.7	1.02
0.90	13.9	1.06	—	—	13.9	1.03
1.00	12.5	1.07	12.5	1.01	12.3	1.03
1.25	9.96	1.10	10.0	1.02	9.98	1.04
1.50	8.30	1.14	8.34	1.03	8.31	1.04
1.75	7.11	1.19	7.14	1.04	7.17	1.04
2.00	6.22	1.24	6.26	1.05	6.35	1.05
2.25	5.59	1.28	5.56	1.06	—	—
2.50	5.03	1.33	5.01	1.08	5.05	1.07
3.00	4.19	1.43	4.17	1.11	4.13	1.11
3.50	3.62	1.52	3.58	1.15	3.53	1.11
4.00	3.17	1.61	3.13	1.18	3.16	1.12
4.50	2.85	1.69	—	—	2.91	1.12
5.00	2.56	1.77	—	—	2.58	1.12
Aluminum Conductors						
0.40	51.2	1.01*	—	—	—	—
0.50	41.0	1.01	—	—	—	—
0.60	34.2	1.01	—	—	—	—
0.70	29.3	1.01	—	—	—	—
0.75	27.3	1.01	—	—	27.9	1.01
0.80	25.6	1.02	—	—	25.8	1.01
0.90	22.8	1.02	—	—	22.8	1.01
1.00	20.5	1.03	20.6	1.01	20.2	1.01
1.25	16.4	1.04	16.5	1.01	16.4	1.01
1.50	13.7	1.06	13.7	1.01	13.7	1.01
1.75	11.7	1.08	11.8	1.02	11.8	1.02
2.00	10.3	1.10	10.3	1.02	10.5	1.02
2.25	9.19	1.12	9.15	1.03	—	—
2.50	8.28	1.14	8.24	1.03	8.31	1.03
3.00	6.89	1.20	6.86	1.04	6.80	1.04
3.50	5.96	1.25	5.89	1.06	5.81	1.04
4.00	5.22	1.31	5.15	1.08	5.20	1.05
4.50	4.69	1.36	—	—	4.78	1.05
5.00	4.22	1.42	—	—	4.25	1.05

Table 2.2.2. (continued)
DC RESISTANCE, AC/DC RESISTANCE RATIO OF CABLE CONDUCTORS

Approximate magnitude of conductor proximity effect on single conductor cable — with each cable installed in a separate nonmetallic duct and operated with open circuited sheath 65 C and 60 cycles

			Axial spacing between adjacent cables (in.)				
			5.0	**6.0**	**7.0**	**8.0**	**9.0**
Cdr Size CI	Cdr Diam. (in.)	Cdr D-C Res.	Conductor proximity effect-percent of DC resistance				

Copper Conductors
Compact Segmental Conductors (Treated)

Cdr Size CI	Cdr Diam. (in.)	Cdr D-C Res.	5.0	6.0	7.0	8.0	9.0
4.00	2.309	3.13	8.7	6.0	4.4	3.4	2.7
3.50	2.159	3.58	6.5	4.5	3.3	2.5	2.0
3.00	1.998	4.17	4.5	3.1	2.3	1.7	1.4
2.50	1.824	5.01	2.8	1.9	1.4	1.1	*
2.00	1.632	6.26	1.5	1.1	*	*	*
1.75	1.526	7.14	1.1	*	*	*	*

Concentric Round Conductors (Dry)

Cdr Size CI	Cdr Diam. (in.)	Cdr D-C Res.	5.0	6.0	7.0	8.0	9.0
4.00	2.309	3.17	18.3	12.5	9.1	7.0	5.5
3.50	2.159	3.62	15.0	10.3	7.5	5.7	4.5
3.00	1.998	4.19	11.9	8.2	6.0	4.6	3.6
2.50	1.824	5.03	8.8	6.1	4.4	3.4	2.7
2.00	1.632	6.22	5.9	4.1	3.0	2.3	1.8
1.75	1.526	7.11	4.6	3.3	2.3	1.8	1.4
1.50	1.412	8.30	3.3	2.3	1.7	1.3	1.0
1.25	1.289	9.96	2.2	1.5	1.1	*	*
1.00	1.152	12.5	1.3	*	*	*	*

Aluminum Conductors
Compact Segmental Conductors (Treated)

Cdr Size CI	Cdr Diam. (in.)	Cdr D-C Res.	5.0	6.0	7.0	8.0	9.0
4.00	2.309	5.15	4.3	3.0	2.2	1.7	1.3
3.50	2.159	5.89	3.0	2.1	1.5	1.2	*
3.00	1.998	6.86	2.0	1.4	1.0	*	*
2.50	1.824	8.24	1.2	*	*	*	*

Concentric Round Conductors (Dry)

Cdr Size CI	Cdr Diam. (in.)	Cdr D-C Res.	5.0	6.0	7.0	8.0	9.0
4.00	2.309	5.22	13.9	9.6	7.0	5.3	4.2
3.50	2.159	5.96	10.9	7.5	5.5	4.2	3.3
3.00	1.998	6.89	8.1	5.6	4.1	3.1	2.5
2.50	1.824	8.28	5.5	3.8	2.8	2.1	1.7
2.00	1.632	10.3	3.4	2.3	1.7	1.3	1.0
1.75	1.526	11.7	2.5	1.7	1.2	1.0	*
1.50	1.412	13.7	1.7	1.1	*	*	*
1.25	1.289	16.4	1.0	*	*	*	*
1.00	1.152	20.5	*	*	*	*	*

Note: Asterisk * indicates the effect is less than 1%. DC resistances are given in microohms per foot of conductor at 65 °C.

Table 2.2.3.
VALUES OF K_1

Conductor temperature (°C)	Temperature factor k_1	
	Copper conductor	Aluminum conductor
20	1.000	1.000
50	1.118	1.121
55	1.138	1.141
60	1.157	1.161
65	1.177	1.181
70	1.197	1.202
75	1.216	1.222
80	1.236	1.242
85	1.255	1.262
90	1.275	1.282
95	1.295	1.302

Simplified form when $x < 2.8$

$$\lambda_s = F(x) = \frac{X^4}{192 + 0.8\, X^4} \tag{2.2.6.}$$

$$X = \frac{8\,\pi + \mu_s\, K_{s1}}{\gamma_0\, k_1 \times 10^9} \tag{2.2.7.}$$

where $K_{s1} = 1$ (nonsegmented conductor), 0.44 (4-segmented conductor), 0.39 (6-segmented conductor), $\lambda_1 k_1 = $ DC resistivity (Ω-cm) at the operating temperature, $\mu_s = $ the specific permeability ($\mu_s = 1.0$ for Cu and Al), and $f = $ frequency.

$$\lambda_p = \frac{\dfrac{3}{2}\left(\dfrac{d_1}{S}\right)^2 G(X')}{1 - \dfrac{5}{24}\left(\dfrac{d_1}{S}\right)^2 H(X')} \tag{2.2.8.}$$

and a simplified form when $X' < 2.8$

$$\lambda_p = \frac{X'^4}{192 + 0.8 X'^2}\left(\frac{d_1}{S}\right)^2 \times \left\{ 0.312\left(\frac{d_1}{S}\right)^2 + \frac{1.18}{\dfrac{X'^4}{192 + 0.8 X'^4} + 0.27} \right\} \tag{2.2.9.}$$

where $d_1 = $ the outer conductor diameter, $S = $ inter-conductor center-to-center) distance, $X' = 0.894X$

$$G(X') = \frac{X'}{4}\; \frac{\text{ber } X'\,\text{ber}'\, X' + \text{bei } X'\,\text{bei}'\, X'}{(\text{bei } X')^2 + (\text{ber } X')^2} \tag{2.2.10.}$$

$$H(X') = \frac{F(X')}{G(X')} \tag{2.2.11.}$$

Sector-Shaped Conductor

λ_s is the same as for a round conductor.

$$\lambda_p = \frac{\frac{5}{4}\left(\frac{d_1}{S}\right)^2 G(X')}{1 - \frac{5}{24}\left(\frac{d_1}{S}\right)^2 H(X')}$$ (2.2.12.)

where (d_1/S) may be equivalent to that of round conductor.

Hollow Conductor

$$X = \sqrt{\frac{8\pi f \mu_s K_{s2}}{\tau_0 k_1 \times 10^9}}$$ (2.2.13.)

$$K_{s2} = \frac{d_1 - d_0}{d_1 + d_0}\left(\frac{d_1 + 2d_0}{d_1 + d_0}\right)^2$$ (2.2.14.)

where d_1 = the outer conductor-diameter and d_0 = the inner conductor-diameter. Values of $F(X)$ and $G(X)$ are tabulated in Table 2.2.4. and Table 2.2.5., respectively.[3]

D. Measured Values of AC/DC Resistance Ratio for Large Stranded Conductors

It is still difficult to make correct or exact estimate of the AC/DC resistance ratio of a specific cable system with large size conductors because its circumferential conditions are usually undetermined. It is therefore necessary to measure the ratio by simulating actual conditions or *in situ*. According to Reference 25, the measurement of the AC/DC resistance ratio was carried out for practical reasons on a 550 kV paper-insulated pipe cable with 2000 MCM conductors, under the assumption that:

1.   The resistance does not change with time — although the interstrand, inter-segment, inter-tape, and tape-skid wire resistances may change
2.   The ratio is a function only of conductor temperature, whereas the temperatures of the shield and pipe, and hence the thermal gradient, are factors
3.   The ratio is independent of current — but if, during measurement, the current is adjusted to give temperatures of interest, the effects of current and temperature are inextricably coupled
4.   The ratio is constant — although as the cables change position in the pipe it is known that the ratio changes

AC resistance may be measured with an AC resistance bridge, a low power factor watt meter, and sometimes an AC volt meter. More than two methods may be employed as a mutual check on accuracy. Some experimental values of AC/DC resistance ratio for 500 kV POF cables are shown in Table 2.2.6.[10,25,26]

E. Conductor Shape

Conductors are generally stranded in order to give enough flexibility to comparatively rigid copper and aluminum conductors. For aluminum conductor cable, solid-type conductor has also been put to practical use with the hope of reducing cost and the outer diameter of the cable. While most single-conductor cables are of the concentric-strand type, they may also be compact-round, annular-stranded, segmental, or hollow-core.[1,2]

Compact or compressed conductors have received wide acceptance recently. The principal advantages of such conductors are reduced overall diameter for a given conductor cross

Table 2.2.4.
F (x)

	0	1	2	3	4	5	6	7	8	9
0.3	—	—	—	—	—	—	—	—	—	—
0.4	—	—	—	—	—	—	—	—	—	—
0.5	—	—	—	—	0.001	0.001	0.001	0.001	0.001	0.001
0.6	0.001	0.001	0.001	0.001	0.001	0.001	0.001	0.001	0.001	0.001
0.7	0.001	0.001	0.001	0.002	0.002	0.002	0.002	0.002	0.002	0.002
0.8	0.002	0.002	0.002	0.003	0.003	0.003	0.003	0.003	0.003	0.003
0.9	0.003	0.004	0.004	0.004	0.004	0.004	0.005	0.005	0.005	0.005
1.0	0.005	0.005	0.006	0.006	0.006	0.007	0.007	0.007	0.007	0.007
1.1	0.008	0.008	0.008	0.008	0.009	0.009	0.010	0.010	0.010	0.010
1.2	0.011	0.011	0.011	0.012	0.012	0.013	0.013	0.013	0.014	0.014
1.3	0.015	0.015	0.016	0.016	0.017	0.017	0.018	0.018	0.019	0.019
1.4	0.020	0.020	0.021	0.021	0.022	0.023	0.023	0.024	0.025	0.025
1.5	0.026	0.027	0.027	0.028	0.029	0.029	0.030	0.031	0.032	0.032
1.6	0.033	0.034	0.035	0.036	0.037	0.038	0.038	0.039	0.040	0.041
1.7	0.042	0.043	0.044	0.045	0.046	0.047	0.048	0.049	0.050	0.051
1.8	0.052	0.054	0.055	0.056	0.057	0.058	0.059	0.061	0.062	0.063
1.9	0.064	0.066	0.067	0.068	0.070	0.071	0.072	0.074	0.075	0.077
2.0	0.078	0.080	0.081	0.083	0.084	0.086	0.087	0.089	0.091	0.092
2.1	0.094	0.095	0.097	0.099	0.101	0.102	0.104	0.106	0.108	0.109
2.2	0.111	0.113	0.115	0.117	0.119	0.121	0.123	0.125	0.127	0.129
2.3	0.131	0.133	0.135	0.137	0.139	0.141	0.143	0.145	0.148	0.150
2.4	0.152	0.154	0.157	0.159	0.161	0.164	0.166	0.168	0.172	0.173
2.5	0.175	0.178	0.180	0.183	0.185	0.188	0.190	0.193	0.196	0.198
2.6	0.201	0.203	0.206	0.209	0.211	0.214	0.217	0.219	0.222	0.225
2.7	0.228	0.230	0.233	0.236	0.239	0.242	0.245	0.247	0.250	0.253
2.8	0.256	0.259	0.262	0.265	0.268	0.271	0.274	0.277	0.280	0.283
2.9	0.287	0.290	0.293	0.296	0.299	0.302	0.305	0.309	0.312	0.315
3.0	0.318	0.321	0.325	0.328	0.331	0.334	0.338	0.341	0.344	0.348
3.1	0.351	0.354	0.358	0.361	0.365	0.368	0.371	0.375	0.378	0.382
3.2	0.385	0.389	0.392	0.396	0.399	0.402	0.406	0.409	0.413	0.417
3.3	0.420	0.424	0.427	0.431	0.434	0.438	0.441	0.445	0.449	0.452
3.4	0.456	0.459	0.463	0.467	0.470	0.474	0.477	0.481	0.485	0.488
3.5	0.492	0.496	0.499	9.503	0.507	0.510	0.514	0.518	0.521	0.525
3.6	0.529	0.533	0.536	0.540	0.544	0.547	0.551	0.555	0.559	0.562
3.7	0.566	0.570	0.573	0.577	0.581	0.585	0.588	0.592	0.596	0.599
3.8	0.603	0.607	0.611	0.614	0.618	0.622	0.626	0.629	0.633	0.637
3.9	0.641	0.644	0.648	0.652	0.656	0.659	0.663	0.667	0.670	0.674
4.0	0.678	0.682	0.685	0.689	0.693	0.697	0.700	0.704	0.708	0.711
4.1	0.715	0.719	0.723	0.726	0.730	0.734	0.738	0.741	0.745	0.749
4.2	0.752	0.756	0.760	0.763	0.767	0.771	0.775	0.778	0.782	0.786
4.3	0.789	0.793	0.797	0.804	0.804	0.808	0.811	0.815	0.819	0.823
4.4	0.826	0.830	0.834	0.836	0.841	0.845	0.848	0.852	0.856	0.859
4.5	0.863	0.866	0.870	0.874	0.877	0.881	0.885	0.888	0.892	0.896
4.6	0.899	0.903	0.906	0.910	0.914	0.917	0.921	0.925	0.928	0.932
4.7	0.935	0.939	0.943	0.946	0.950	0.953	0.957	0.961	0.964	0.968
4.8	0.971	0.975	0.979	0.982	0.986	0.989	0.993	0.996	1.000	1.004
4.9	1.007	1.011	1.014	1.018	1.021	1.025	1.029	1.032	1.036	1.039
5.0	1.043	—	—	—	—	—	—	—	—	—

section; elimination of space between the conductor and the insulation, which results in higher electrical breakdown; low AC resistance due to minimizing of proximity effect; retention of the close stranding during bending; and for solid cables, elimination of many longitudinal channels along which impregnating compound can migrate. To achieve minimum diameter by conductor processing, the strands of a concentric-round conductor are formed by rolling each layer before the next layer is applied.

Table 2.2.5.
G (x)

	0	1	2	3	4	5	6	7	8	9
0.3	—	—	—	—	—	—	—	—	—	—
0.4	—	—	—	—	—	—	—	—	—	—
0.5	0.001	0.001	0.001	0.001	0.001	0.002	0.002	0.002	0.002	0.002
0.6	0.002	0.002	0.002	0.002	0.002	0.002	0.003	0.003	0.003	0.003
0.7	0.004	0.004	0.004	0.004	0.004	0.004	0.005	0.005	0.005	0.005
0.8	0.006	0.006	0.006	0.006	0.007	0.007	0.008	0.008	0.009	0.009
0.9	0.010	0.010	0.010	0.011	0.011	0.012	0.012	0.013	0.014	0.014
1.0	0.015	0.015	0.016	0.017	0.017	0.018	0.019	0.019	0.020	0.021
1.1	0.022	0.023	0.023	0.024	0.025	0.026	0.027	0.028	0.029	0.030
1.2	0.031	0.032	0.033	0.034	0.035	0.036	0.037	0.038	0.039	0.040
1.3	0.041	0.043	0.044	0.046	0.046	0.047	0.049	0.050	0.052	0.053
1.4	0.054	0.055	0.057	0.058	0.060	0.061	0.063	0.064	0.066	0.067
1.5	0.069	0.070	0.072	0.073	0.075	0.076	0.078	0.080	0.082	0.084
1.6	0.086	0.087	0.089	0.091	0.093	0.095	0.097	0.099	0.100	0.104
1.7	0.106	0.107	0.109	0.110	0.113	0.115	0.117	0.119	0.120	0.124
1.8	0.126	0.129	0.130	0.132	0.133	0.135	0.140	0.141	0.144	0.147
1.9	0.149	0.151	0.152	0.153	0.158	0.160	0.162	0.165	0.168	0.170
2.0	0.172	0.175	0.177	0.179	0.182	0.184	0.186	0.188	0.190	0.193
2.1	0.197	0.198	0.200	0.204	0.206	0.209	0.212	0.214	0.216	0.219
2.2	0.221	0.224	0.226	0.229	0.232	0.234	0.236	0.238	0.242	0.244
2.3	0.246	0.249	0.251	0.253	0.256	0.259	0.262	0.263	0.266	0.269
2.4	0.271	0.273	0.275	0.278	0.280	0.282	0.285	0.287	0.290	0.292
2.5	0.295	0.297	0.298	0.300	0.302	0.305	0.308	0.311	0.313	0.315
2.6	0.318	0.320	0.322	0.324	0.326	0.329	0.331	0.333	0.336	0.339
2.7	0.341	0.343	0.344	0.346	0.348	0.351	0.353	0.356	0.358	0.360
2.8	0.363	0.365	0.367	0.369	0.371	0.373	0.375	0.377	0.379	0.382
2.9	0.384	0.386	0.388	0.390	0.392	0.394	0.396	0.398	0.400	0.402
3.0	0.405	0.407	0.409	0.411	0.413	0.415	0.417	0.418	0.420	0.422
3.1	0.425	0.426	0.428	0.430	0.432	0.433	0.435	0.437	0.440	0.441
3.2	0.444	0.445	0.447	0.449	0.451	0.453	0.455	0.457	0.459	0.461
3.3	0.463	0.464	0.466	0.467	0.470	0.472	0.474	0.475	0.477	0.479
3.4	0.481	0.483	0.484	0.486	0.488	0.490	0.492	0.494	0.496	0.497
3.5	0.499	0.501	0.502	0.504	0.506	0.508	0.510	0.511	0.512	0.514
3.6	0.516	0.518	0.520	0.521	0.521	0.525	0.526	0.528	0.530	0.531
3.7	0.533	0.535	0.536	0.538	0.540	0.542	0.544	0.545	0.547	0.549
3.8	0.550	0.552	0.554	0.555	0.557	0.558	0.560	0.562	0.564	0.565
3.9	0.567	0.569	0.570	0.572	0.574	0.576	0.577	0.579	0.580	0.582
4.0	0.584	0.585	0.587	0.589	0.591	0.592	0.594	0.596	0.598	0.599
4.1	0.601	0.602	0.604	0.606	0.608	0.610	0.612	0.613	0.615	0.616
4.2	0.618	0.620	0.621	0.623	0.625	0.627	0.629	0.630	0.632	0.634
4.3	0.635	0.637	0.639	0.640	0.642	0.644	0.645	0.647	0.650	0.651
4.4	0.652	0.654	0.655	0.657	0.659	0.661	0.662	0.664	0.666	0.667
4.5	0.669	0.671	0.673	0.674	0.676	0.678	0.680	0.681	0.683	0.685
4.6	0.686	0.686	0.690	0.691	0.693	0.695	0.697	0.698	0.700	0.702
4.7	0.703	0.705	0.707	0.709	0.710	0.712	0.714	0.716	0.718	0.719
4.8	0.720	0.721	0.724	0.726	0.728	0.730	0.731	0.733	0.735	0.736
4.9	0.738	0.740	0.741	0.743	0.744	0.745	0.748	0.750	0.752	0.753
5.0	0.755	—	—	—	—	—	—	—	—	—

Table 2.2.7. shows a classification of conductor shapes. The segmental type is preferable for large-size conductor, say 600 to 800 mm^2 (1000 kcmil to 1300 kcmil) or larger, in which the skin effect is not negligible. A segmental, compact, round, stranded conductor comprising several compacted segments insulated from each other by insulating paper and the like, has been widely used. Single-core OF cable is equipped with an oil path in the center of its conductor. This type of conductor, which is called a hollow core, round, stranded conductor, is less influenced by skin effect than is the stranded, round conductor.

Table 2.2.6.
AC/DC RESISTANCE RATIO OF 500 kV POF CABLES

Measurement method	Conductor temperature (°C)	Power current (A)	AC/DC Resistance ratio	Remarks
Uncertain	20		1.64[c]	2000 MCM compact
	40		1.58[c]	Segmental-coated copper
	60		1.53[c]	Conductor, 4-segments
	80		1.47[c]	2-Insulated
	100		1.41[c]	40-ft length of 12-3/4 in. O.D. pipe
AC volt meter[a]	44	525	1.27 ⎫	10-3/4 in. pipe
	44	863	1.18 ⎭	
	36	525	1.50 ⎫	12-3/4 in. pipe
	36	900	1.32 ⎭	
				2000 MCM, tinned-copper 5 segments, 3 nylon tape insulated
AC resistance bridge[b]	25	800	1.79	2000 MCM compact
	50	800	1.70	Segmental-tinned strands
	77	800	1.64	3—0.006 in. Paper tapes, butted on opposite segments

[a] Reference 25.
[b] Reference 26.
[c] Readings from a figure in Reference 10.

Table 2.2.7.
CONDUCTOR SHAPE

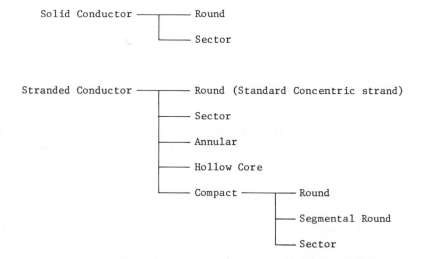

Because of the conductor losses, the largest size of copper conductor that can be used for POF cable is 2500 kcmil or 1600 mm². The pipe loss and the AC resistance of the conductors may be reduced by using stainless steel pipe.[37] When these losses are reduced, it is possible to use copper conductors larger than 2500 kcmil (1600 mm²); sizes up to 3500 kcmil or 2200 mm² become feasible. The AC/DC resistance ratio for such a larger copper conductor can be reduced by the use of enamel on the individual strands. Aluminum conductors up to 3500 kcmil can be used because of the high contact resistance between strands, due to the oxide coating, which reduces the AC resistance.

2.2.2. Insulating Materials

A variety of insulating materials are used in cable manufacture. These include natural and synthetic rubber (thermo-setting) compounds, varnished cambric, impregnated paper, polyvinyl chloride and polyethylene (thermoplastics), and gases. Factors influencing choice are rated voltage and current and the type of cable. The use of oil-impregnated paper has shifted toward the higher voltage class, while extruded thermoplastics have superceded paper in the lower voltage class. PE and XLPE insulation has undergone a dramatic technical development for higher voltage (even EHV) applications. Polyvinyl chloride is limited to low-voltage power and control circuit cables.

The development of new types of cable, such as cryoresistive cable and superconducting cable, has forced exploration of new insulating materials which can operate at very low temperatures. These may be plastic, liquid, gaseous, or vacuum. Installation of many compressed-gas insulated substations has stimulated the development of compressed-gas insulated cable. At the present time, sulfurhexafluoride (SF_6) is the principal gas used but other gases are under investigation.

DC insulation seems to be totally different from AC insulation, a fact that was not recognized until recently. Oil-impregnated paper works more or less satisfactorily, but XLPE and PE do not. The reason for this is not yet clear.

A. Characteristics Required

General performance characteristics that any AC cable insulation should possess can be itemized as follows:[3]

1. It should withstand high AC and impulse voltages and should remain stable in this respect for a reasonably long period of time.
2. Its dielectric power loss should be as small as possible.
3. It should be excellent in treeing resistance and corona resistance.
4. It should be flexible and antiabrasive.

B. Oil-Impregnated Paper

Oil-impregnated paper is literally a composite of insulating paper and oil; it is the most common form of cable insulation used for bulk transmission and distribution of power. Impregnated-paper insulation consists of multiple layers of paper tapes, each tape from 2.5 mil (50 μm) to 7.5 mil (200 μm) in thickness, wrapped helically around the conductor to be insulated. The resulting insulation is comparatively homogeneous. One of the most important features of oil-impregnated paper is its high resistance to and/or freedom from corona or internal discharges. This provides the stability needed in cable insulation, especially for solid impregnated cables, and it is a main reason for its use in such applications.

Unfortunately, its electrical performance can be easily affected by a small amount of water. Figure 2.2.1.[52,53] shows the variation of tanδ due to the addition of water, measured in parts per million (ppm). Thermal degradation is also enhanced by a small amount of water. For this reason it is a standard practice to dry paper in vacuum before oil impregnation and to protect a finished cable from water permeation with a lead sheath.

The quality of the impregnated-paper insulation depends not only on the properties and characteristics of the paper and impregnating oil used, but even more on the mechanical application of the paper tapes to the conductor, the thoroughness of the vacuum drying, and the control of the saturating and cooling cycle during impregnation.

Table 2.2.8. shows characteristics of several kinds of insulating paper presently used for power cables.[3] Generally speaking, paper with a density less than 0.75 g/cm³ is called the low-density paper, and paper with more than 0.9 g/cm³ is high-density paper. Thin (thinner than 100 μm) and thick (thicker than 150 μm) papers are available for power cable use.

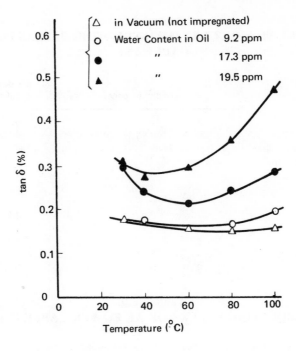

FIGURE 2.2.1.   Effects of water content on loss tangent of oil-impregnated paper. (From Tanaka, T. and Fukuda, T., Tech. Rep. No. 70005, Central Research Inst. of Electric Power Industry, Tokyo, 1970, 14.)

Paper thickness is determined by voltage rating, conductor diameter, and other requirements; the paper layers may be the same thickness or they may be graded. Table 2.2.9. shows several kinds of insulating paper for impregnated-paper insulated cables[3] classified according to service voltage.

The insulation characteristics of impregnated paper are affected by a number of factors such as impermeability or airtightness, density, thickness, impurities, and the presence of water.

C. Effects of Impermeability

Impermeability is a measure of the fineness of fibers and their homogeneity; it is thought to affect the breakdown voltage of the paper. It is closely related to the density of the paper, so much so that the effects of the two factors are practically inseparable. As a consequence, there is controversy over the effect of the impermeability on breakdown voltage. It is at least agreed among investigators that impulse breakdown voltage increases as the impermeability increases up to about 10^7 in Emanueli units (e.u.); (e.u. = [Gurley (sec/100 cc) per thickness (mm)] × 460). AC and DC breakdown voltages may also be positive functions of impermeability up to a certain value.

D. Effects of Density

The relation between breakdown voltage and paper density has not been established with certainty. For a small model cable, it has been demonstrated that the AC and impulse breakdown voltages are almost independent of the density below 0.9 g/cm^3, and decrease slightly above 0.9 g/cm^3.[3] This is interpreted in terms of the increase in dielectric constant of the paper, which is the fraction of the voltage sustained by the oil.

Theoretically and experimentally, evidence supports the fact that as the density of paper decreases, the dielectric constant and dissipation factor decrease; this is shown in Figure

Table 2.2.8.

CHARACTERISTICS OF CERTAIN INSULATING PAPERS FOR POWER CABLE USE

Item	Ordinary paper				Low density paper washed by deionized water		
Thickness (μm)	25	40	70	125	70	125	200
Density (g/cm³)	0.95	0.96	0.92	0.85	0.71	0.72	0.66
Impermeability, Gurley (sec/100ᶜᶜ)	11,000	5,000	1,300	620	6,100	1,800	340
Tensile strength (kg/15 mm)							
Longitudinal	4.0	6.1	15.2	13.0	7.4	11.8	15.5
Transverse	2.5	4.2	4.0	5.9	4.5	5.8	8.5
Elongation							
Longitudinal	2.0	2.0	2.9	2.5	2.2	2.3	2.5
Transverse	5.2	6.2	9.8	7.0	5.1	6.1	4.9
Oil-impregnated paper							
Specific dielectric constant	3.82	3.83	3.65	3.53	3.40	3.42	3.35
tanδ (%) at 80°C	0.20	0.21	0.24	0.22	0.16	0.17	0.17

Table 2.2.9.

INSULATING PAPER FOR AC POWER CABLE USE

Operating voltage (kV)		Thickness (μm)	Density (g/cm³)	Impermeability (sec/100 cc)	Remarks
22		125	@ 0.85	500 ∼ 1000	—
66 ∼ 77					
154	OF	70 ∼ 200	0.7 ∼ 0.95	500 ∼ 1000	Deionized paper grading
	POF				
275	OF	70 ∼ 200	0.7 ∼ 0.95	500 ∼ 1000	Deionized paper grading
	POF				
500	OF	70 ∼ 200	0.65 ∼ 0.95	500 ∼ 1000	Deionized paper grading
	POF				

2.2.2.[54] It would appear that this is a way to improve the dielectric performance of paper. Unfortunately, contrary to the results obtained from the model cable experiment, actual cables, consisting of low-density paper, show a decrease in breakdown voltage. This may be due to the reduction in mechanical strength of the paper. Paper with low mechanical strength is liable to wrinkle and crease when the cable is handled. It is well known that wrinkles and creases reduce the breakdown voltage.

Improvements in paper and cable manufacturing techniques have been instituted which tend to reduce mechanical disadvantage for EHV and UHV cables. It remains a developmental objective to manufacture paper with low density and high impermeability, although the two attributes appear to be inconsistent.

E. Effects of Thickness

Reduction in paper thickness reacts favorably on the breakdown stress. This arises from the resulting reduction in oil-gap length, increase in barrier effect, and in impermeability. The barrier effect is caused by the arrangement of the stacked fine fiber. Figure 2.2.3. shows a linear relationship between impulse breakdown stress and the logarithm of paper thickness.[54] Figure 2.2.4. shows a similar dependence for AC voltage breakdown.[54]

In actual application, the advantage of thin paper will be diminished to a certain degree by the disadvantage of mechanical strength and size, consequently there would appear to be an optimum thickness. Nowadays, 70 μm or greater is generally used, although paper as thin as 25 μm or 40 μm has been used on a trial basis.

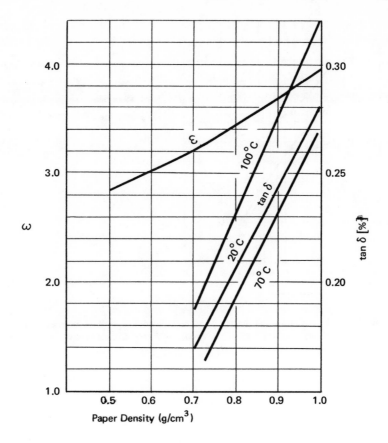

FIGURE 2.2.2. Dependence of permittivity and loss tangent on density of oil-impregnated paper. (From Saito, Y. and Take, Y., *Electrical Insulating Paper*. Corona. Tokyo, Japan. 1969, 431. With permission.)

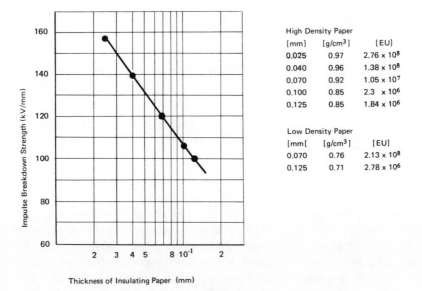

High Density Paper		
[mm]	[g/cm^3]	[EU]
0.025	0.97	2.76 x 10^8
0.040	0.96	1.38 x 10^8
0.070	0.92	1.05 x 10^7
0.100	0.85	2.3 x 10^6
0.125	0.85	1.84 x 10^6

Low Density Paper		
[mm[	[g/cm^3]	[EU]
0.070	0.76	2.13 x 10^8
0.125	0.71	2.78 x 10^6

FIGURE 2.2.3. Thickness dependence of impulse breakdown strength of high and low density oil-impregnated paper (model cable experiment). (From Saito, Y. and Take, Y., *Electrical Insulating Paper*, Corona, Tokyo, Japan, 1969, 431. With permission.)

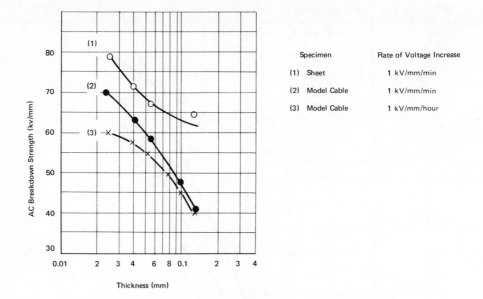

FIGURE 2.2.4.   Thickness dependence of AC breakdown strength of oil-impregnated paper. (From Saito, Y. and Take, Y., *Electrical Insulating Paper*, Corona, Tokyo, Japan, 1969, 433. With permission.)

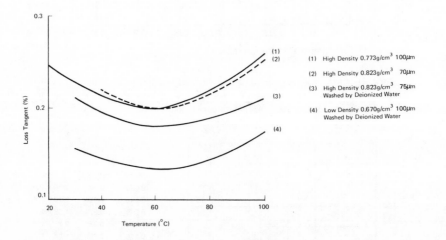

FIGURE 2.2.5.   Temperature dependence of loss tangent of cellulose paper available for oil-filled type cables. (From Iizuka, K., **Ed.,** *Power Cable Technology Hand Book*, Denki-Shoin, Tokyo, 1974, 80. With permission.)

F. Dielectric Properties of Paper

The dielectric properties of oil-impregnated paper are much dependent on the content of water absorbed in the paper; they also depend on the density of the paper and a few ionic impurities. Generally, the effect of less than 0.1% of water on tanδ is negligible for all practical purposes.[54] Figure 2.2.5. shows the effects of density and ionic impurity on tanδ.[3] For EHV cable, low-density, deionized water-washed paper is utilized because of its low dielectric loss. The dissipation factor tends to increase as insulation degradation proceeds. In case of solid-type, impregnated-paper insulated cable, this phenomenon is well interpreted in terms of corona attack on the insulation. It is more complicated for OF and POF cables. Several probable mechanisms for the degradation are

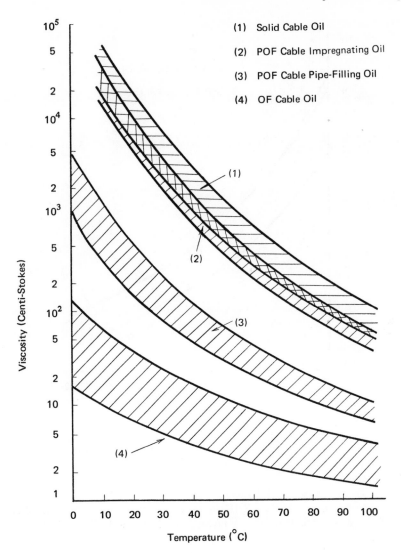

(1) Solid Cable Oil

(2) POF Cable Impregnating Oil

(3) POF Cable Pipe-Filling Oil

(4) OF Cable Oil

FIGURE 2.2.6. Viscosity criteria of cable oil for three types of OF cables. (From Iizuka, K., **Ed.,** *Power Cable Technology Hand Book,* Denki-Shoin, Tokyo, 1974, 82. With permission.)

1. The decrease in the degree of polymerization of paper due to heating
2. The dissolving-out of ionic impurities from the paper into the oil
3. The acceleration of thermal degradation in the oil by the catalytic reaction of copper

In spite of these factors, it is certain that oil-impregnated paper insulation is stable as long as oxygen and water are excluded and it is subjected to temperatures less than 80 to 85°C.

G. Insulating Oil for OF Cable Use

Insulating oil for power cable use should retain excellent electrical properties for a long period of time and also possess the correct viscosity. Figure 2.2.6. shows a standard criterion for insulating oils used in several kinds of OF-type cables. Insulating oil should have a low viscosity for OF cable use.[3] Two types of oils are available: mineral oil and alkylbenzene synthetic oil. Figure 2.2.7. shows the temperature dependence of viscosity for three kinds

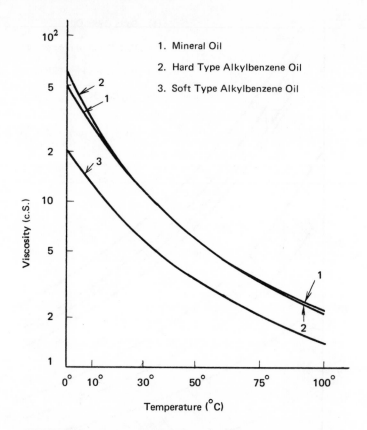

FIGURE 2.2.7.   Temperature dependence of viscosity of three kinds of oil.
(From Iizuka, K., **Ed.,** *Power Cable Technology Hand Book,* Denki-Shoin,
Tokyo, 1974, 82. With permission.)

of oil.[3] Figure 2.2.8. shows their hydrogen gas absorption properties which are closely
related to corona resistance.[3] Table 2.2.10. tabulates various performance characteristics of
oil used in OF cable.[3] Alkylbenzene is favored for EHV or UHV cable because of its
excellent gas absorption characteristic and its lack of chemical reactivity with copper.

H. Insulating Oil for POF Cable Use

Impregnating oil for POF cable should be sufficiently viscous to assure that it does not
migrate prior to cable installation. Pipe filling oil should have as low a viscosity as possible
as far as flow resistance requirements are concerned, but it might better be very viscous in
the event of an oil leak accident. Mineral oil and polybutane synthetic oil are in wide use
for both insulation impregnation and pipe filling. Figure 2.2.9. shows the temperature
dependence of the viscosity of oils used for POF cable.[3] Table 2.2.11. tabulates their various
characteristics.[3]

I. Oil for Solid Type and Other Cables

Oil draining is a principal concern for SL cables, belted cables, gas-filled OF cables, and
pipe-type gas compression OF cables. Insulating oil should be as viscous as the manufacturing
process will permit. Nondraining compound is used for highly uneven cable installations.
Table 2.2.12. shows some characteristics of oils used in solid cable.[3]

J. Synthetic Paper for UHV Cable Insulation

Many kinds of materials have been developed to meet such UHV cable insulation requirements as a decrease in $\epsilon \tan \delta$ and an increase in breakdown stress. The low-density

(1) Alkylbenzenc Synthetic Oil

(2) OF Cable Use Mineral Oil

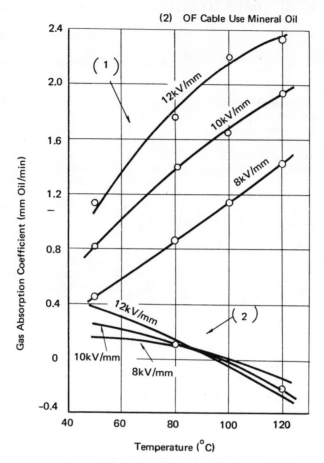

FIGURE 2.2.8. Hydrogen gas absorption characteristics of two kinds of oil. (From Iizuka, K. **Ed.,** *Power Cable Technology Hand Book,* Denki-Shoin, Tokyo, 1974, 83. With permission.)

deionized paper presently available may not be applicable for nonforced-cooled UHV cable. Improved papers have been developed and are available in such forms as stuck-together paper, laminated paper, polyethylene-loaded paper, mica-loaded paper, glass-flake-contained paper, DN paper, and grafted paper.[70] But reducing the dielectric constant to 3.0 and the dissipation factor to below 0.1% is most difficult. Following Vermeer's proposal,[55,56] plastic films have been intensively investigated. Regrettably, these have problems due to their incompatibility with any impregnating oil because of swelling and crazing, the difficulty of oil impregnation, and the reduction in breakdown voltage by lamination. Various synthetic papers have been developed recently; they have received much attention as substitutes for Kraft® paper in cable insulation. Synthetic and cellulose laminated papers are also interesting. Table 2.2.13.[55-70] shows some characteristics of papers and films under consideration for UHV use.

K. Gas Insulation

Gases for cable insulation should have high breakdown strength, be chemically stable, and be flame retardant. Among many candidates are nitrogen (N_2), sulfurhexafluoride (SF_6), and Freon®-12 (CCl_2F_2). Nitrogen gas is usually selected for gas-filled cables

Table 2.2.10.
TYPICAL CHARACTERISTICS OF OF CABLE USE OILS

| | Mineral oil | Alkylbenzene synthetic oil | |
		Hard type	Soft type
Specific weight (15/4°C)	0.895	0.865	0.864
Viscosity (centistokes, 50°C)	11.9	12.4	5.68
Viscosity (centistokes, 75°C)	3.4	3.3	2.09
Expansion coefficient (1/°C)	0.00075	0.0086	0.0089
Flowage point (°C)	−65	−62.5	−62.5
Ignition point (PMCC) (°C)	136	132	138
Evaporation weight decrease (%)	0.20	0.20	0.15
Reactivity	Neutral	Neutral	Neutral
Total acid value KOH (mg/g)	0.001	0.001	0.002
Corrosiveness	None	None	None
AC breakdown voltage (kV/2.5 mm gap)	55	60	58
$\tan\delta$ (80°C) %			
Original	0.006	0.001	0.001
Degraded[a]	0.102	0.011	0.016
Degraded with copper[a]	8.854	0.010	0.011
Volume resistivity (80°C) Ω-cm			
Original	5×10^{14}	1.2×10^{15}	1.8×10^{15}
Degraded[a]	4×10^{14}	1.6×10^{14}	2.4×10^{14}
Degraded with copper[a]	6×10^{13}	8.9×10^{13}	3.4×10^{14}

[a]   At 115°C for 96 hr.

and gas compression type cables. There have been some attempts to replace N_2 with SF_6 in gas cables already installed to increase service voltage. Some physical and chemical constants of N_2, SF_6, and CCl_2F_2 gases are listed in Table 2.2.14.[3,12,13]

SF_6 is used for compressed gas insulated cable because of its low permittivity, virtually zero dissipation factor, and high breakdown strength. Mixtures of N_2, SF_6, and some other gases are now under investigation.

L. Rubber and Plastic Materials

Extruded cable insulation can be divided into rubber (thermosetting) and plastic (thermoplastic). The former comprises ethylene-propylene rubber and butyl rubber while the latter is polyethylene and cross-linked polyethylene. Various physical, chemical, mechanical, and electrical properties of selected rubber and plastic materials are shown in Table 2.2.15.[15]

M. Polyethylene

Polyethylene is of two kinds, high density and low density. The first is produced by polymerizing ethylene gas into the material at relatively low pressure (1 to 100 atm). The second is polymerized thermally under as high a pressure as 1000 to 2000 atm. High molecular weight polyethylene and low molecular weight polyethylene are virtually synonymous with high- and low-density polyethylene, respectively. There are, however, high molecular weight, low-density polyethylenes and low molecular weight, high-density polyethylenes available. Figure 2.2.10. shows rough quantitative criteria for these parameters.

Polyethylene exhibits an extremely low dissipation factor, high volume resistivity, and high breakdown strength together with excellent resistance to ozone, corona, and weathering. All these characteristics are ideally suited to cable insulation. However, polyethylene is very susceptible to environmental conditions above 75°C and has a critical softening point between 105 and 115°C.

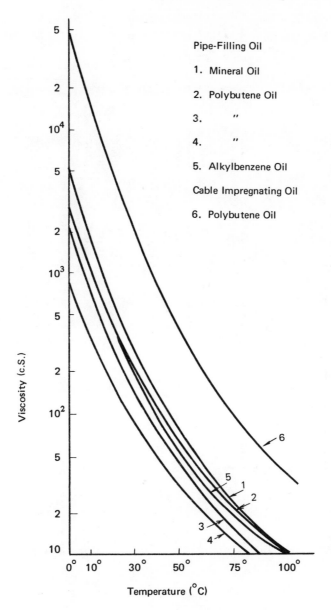

FIGURE 2.2.9.   Temperature dependence of viscosity of POF cable use oils. (From Iizuka, K. **Ed.,** Power Cable Technology Hand Book, Denki-Shoin, Tokyo 1974, 84. With permission.)

N. Cross-Linked Polyethylene

Cross-linked polyethylene is a three-dimensional, net-structured polyethylene with intermolecular bridges. It can be cross linked by irradiation or by chemical reaction. Standard practice for low-density polyethylene is to cross link it with the aid of peroxide, such as DCP (di-αcumyl peroxide) under conditions of high temperature and pressure. Cross linking polyethylene endows it with additional improved performance with respect to thermal deformation, thermal aging, and environmental stress cracking, while retaining almost all the characteristics of permittivity, dissipation factor, and breakdown strength, as compared to regular polyethylene.

Table 2.2.11.
TYPICAL CHARACTERISTICS OF POF CABLE USE OIL

| | Mineral oil 1 | Pipe Filling Oil | | | Alkylbenzene 5 | Impregnating oil Polybutene 6 |
| | | Polybutene | | | | |
		2	3	4		
Specific weight (15/4°C)	0.923	0.850	0.845	0.841	0.876	0.865
Viscosity (C.S.)						
30°C	240	220	142	85.5	189	1800
55°C	53	50	34	25	44	100[b]
Expansion coefficient (1/°C)	0.00072	0.00079	0.00080	0.00080	0.00078	0.00074
Flowage point (°C)	−22.5	−30	−37.5	−35	−32.5	−25
Ignition point (PMCC) (°C)	186	134	142	146	164	156
Evaporation weight decrease (%)	0.07	0.11	0.057	0.12	0.042	0.18
Reactivity	Neutral	Neutral	Neutral	Neutral	Neutral	Neutral
Total acid value KOH (mg/g)	0.002	0.006	0.001	0.007	0.002	0.006
Corrosiveness	None	None	None	None	None	None
AC breakdown voltage (kV/2.5 mm gap)	55	55	50	41	65	45
tanδ (80°C) %						
Original	0.001	0.001	0.001	0.001	0.001	0.001
Degraded[a]	0.046	0.001	0.001	0.001	0.006	0.001
Degraded with copper[a]	0.800	0.002	0.002	0.003	0.010	0.001
Volume Resistivity (80°C) Ω-cm						
Original	5.6×10^{15}	1.0×10^{16}	3.3×10^{15}	1.8×10^{16}	5.7×10^{15}	2.8×10^{15}
Degraded[a]	4.0×10^{13}	2.5×10^{15}	2.1×10^{15}	5.75×10^{15}	4.5×10^{15}	1.4×10^{15}
Degraded with copper[a]	3.6×10^{12}	2.6×10^{15}	2.3×10^{15}	2.1×10^{14}	3.2×10^{15}	1.4×10^{15}

[a] At 115 °C for 96 hr.
[b] A value at 75 °C.

Table 2.2.12.
TYPICAL CHARACTERISTICS OF SOLID CABLE USE OIL

		Ordinary oil	Nondraining compound
Specific weight	15/4°C	0.92	0.92
Viscosity (centistokes)	20°C	8600	—
	40°C	1450	—
	65°C	280	456000
	75°C	163	1200
	80°C	130	500
	100°C	57	88
	120°C	29	46
Expansion coefficient	1/°C	0.0006	0.0006
Flowage point	°C	−10	—
Ignition point (PMCC)	°C	210	—
Reactivity		Neutral	Neutral
Total acid value		0.005	0.006
Corrosiveness		None	None
AC breakdwon voltage (kV/2.5 mm gap)		48 (70°C)	42 (80°C)
tanδ (80°C)%		0.021	0.021
Volume resistivity (80°C) Ω-cm		7.8×10^{13}	7.0×10^{13}

O. Butyl Rubber and EP Rubber

Butyl rubber is a copolymer of isobutylene and isoprene. It has excellent resistance to electric stress, ozone, and heat. It has been superceded by polyethylene and cross-linked polyethylene. EP rubber is a copolymer of ethylene and propylene. It is excellent in its resistance to ozone, thermal aging, weathering, chemicals, and tracking. It is competitive with polyethylene with respect to electric stress. The molecular structures of some of these materials are shown in Figure 2.2.11. and their characteristics are given in Table 2.2.15.

P. Insulating Materials at Cryogenic Temperatures

Possible insulation systems at cryogenic temperatures which may be applicable for cryogenic resistive cables and superconducting cables are

1. Vacuum and spacers
2. Coolant (liquid or gas) and spacers
3. Paper (Kraft® paper or synthetic paper) impregnated with coolant
4. Plastic tape impregnated with coolant
5. Solid plastic cooled by coolant

The insulation systems most likely to be used shall be discussed in a later chapter.

It is necessary to cool a cable system with a suitable coolant, which may play the role of electrical insulator. Table 2.2.16. shows fluid characteristics of various coolants.[72-74] Table 2.2.17. shows some physical constants of coolants and other related materials.[75]

Commercially available polymer films and papers have been tested under various environmental and temperature conditions as shown in Table 2.2.18.,[76] Table 2.2.19.,[77] and Table 2.2.20.[78]

Permittivity and dissipation factors are tabulated in Tables 2.2.21., 2.2.22.,[76] and 2.223. Insulation in paper form is required to enable enough impregnation, especially for liquid nitrogen-cooled cable. Some electrical characteristics are shown in Tables 2.2.24., 2.2.25., and 2.2.26.[79]

Table 2.2.13.

TYPICAL CHARACTERISTICS OF SYNTHETIC PAPER AND OTHERS

Item		Kraft® paper	Tenax	PP	PPLP	PPL	PAP	SSP	POD (film)	FHCP
Thickness (mm)		0.125	0.092	0.164	0.121	0.100	0.125	0.090	0.025	0.128
Density (g/cm³)		0.65	0.77	0.74	0.90	0.90	0.75	0.64	1.38	0.69
Impermeability (s/100 cc)		1,800	1,100	3,200	—	—	3,000	1,600	—	—
Tensile strength (kg/mm²)		5.6	2.3	4.0	5.0	9.4	6.9	7	9	37.8
Elongation (%)		2.6	2.9	15.4	6.2	2.7	25	15	30	10.8
Permittivity (ε)		3.3	2.4	2.2	2.7	2.7	2.65	2.5	3.2	2.3
tan δ (%)	80°C	0.14	0.026	0.02	0.06	0.08	0.045	0.04	0.05	0.024
	100°C	0.15	0.042	0.04	0.073	—	—	0.10	0.12	0.049
Breakdown strength (kV/mm)	AC	74	72	60	—	55	41	70	210	104
	Imp.	156	120	160	172	170	100	125	430	229
$\epsilon \cdot \tan \delta \times 10^{-3}$ (80°C)		4.62	0.63	0.44	1.62	2.16	1.1	1.0	1.6	0.55
Solubility in DDB oil[a] (%)		0.0	0.0	8 (120°C)	2 ~ 3 (100°C)	—	—	—	0.0	0.5 ~ 0.8 (100°C)

[a] DDB: Dodecyl Benzene.

From *J. Inst. Elec. Eng. Jpn.*, 95(5), 403, 1975. With permission.

Table 2.2.14.
PHYSICAL AND CHEMICAL CONSTANTS OF N_2, SF_6, AND CCl_2F_2

	Nitrogen	Sulfurhexafluoride	Freon®-12
Molecular formula	N_2	SF_6	CCl_2F_2
Molecular weight	28.01	146.06	120.9
Density (20°C, 1 atm) (g/ℓ)	—	6.14	—
Relative density to air	1.25	5.07	4.20
Sublimation point (°C)	—	−63.8	—
Melting point (1.24 kg/cm²g)(°C)	−209.86	−50.8	−158
Boiling point (°C)	−195.82	—	−29.8
Critical temperature (°C)	−147.2	45.6	111
Critical pressure (kg/cm²)	33.5	38	41
Critical density (g/cm³)	0.03110	0.725	—
Specific heat (Cp, 30°C, 1 atm) (cal/g°C)	0.492	0.155	—
Heat capacity (30°C, 1 atm) (cal/g · mol · °C)	—	22.6	—
Thermal conductivity (30°C, 1 atm) (cal/s · cm · °C)	5.72×10^{-3}	3.36×10^{-5}	—
Viscosity (30°C, 1 atm) (poise)	1.75×10^{-4} g/cm · s	1.54×10^{-4}	—
Saturated vapor pressure (kg/cm²)			
−40°C		2.65	
−30°C		5.1	1.06
−20°C		6.37	
−15°C		8.4	1.89
0°C		12.4	3.18
20°C		21.4	
40°C		34.5	

2.2.3. Semiconducting Sheath Materials

Electrostatic screens are provided on conductors and insulation to increase the stability of the cable insulation. Carbon paper, metalized paper, and aluminum tape are employed for taped cable; whereas carbon paper, semiconducting cloth (cotton or nylon) tape, semiconducting rubber in the form of a plastic layer, and copper tape are utilized for extruded cable.

A. Carbon Paper[3,80]

Carbon paper screens act to increase appreciably (30 to 50%) the power frequency breakdown voltage and to improve the aging of the insulation. The former is particularly important for the higher voltage range, but the latter is of interest at all voltages. The first result seems to be independent of the kind of carbon paper as long as its volume resistivity is between 10^5 to $10^8 \Omega$-cm. Carbon paper shielding not only eases the electric stress on stranded wires and the shield oil gaps between elemental wires, it also disperses corona discharges and exhibits certain interfacial phenomena.

With thin insulation such as that employed for 33- and 66-kV cables, plain carbon paper screens cause a relatively high increase in power factor which, although not contributing to cable losses, is undesirable for quality control purposes in that it tends to mask possible defects in materials and processing. There are two theories presented to interpret the increase in power factor with applied voltage when carbon paper screens are utilized: one states that the power factor will increase because the resistance in the interface between impregnated paper and carbon paper increases due to the electric stress on the surface of the carbon paper.[81] The other asserts that it increases because the electric field ionizes molecular impurities in the oil which have been absorbed by the carbon paper as a consequence of its inherent absorption capability.[82]

Table 2.2.15.
VARIOUS CHARACTERISTICS OF SELECTED RUBBERS AND PLASTICS

Part A

Plastics properties chart	A.S.T.M. test method	Polyethylenes and ethylene copolymers								PVC
		Low density	Medium density	High density	Polyethylene cross-linkable compounds		Ethylene-ethyl acrylate copolymer	Ethylene vinyl acetate copolymer	High molecular weight	Chlorinated polyvinyl chloride compound
					Molding grades	Wire and cable grades				
Processing										
1. Molding qualities	—	Excellent	Excellent	Excellent	Excellent	—	Excellent	Excellent	Fair	Good
2. Compression molding temp., °F	—	275-350	300-375	300-450	240-450	—	200-300	200-300	400-500	350-400
3. Comp. molding pressure, psi	—	100-800	100-800	500-800	100-800	—	50-1000	50-2500	800-1200	1500-2000
4. Injection molding temp., °F	—	300-500	300-500	300-600	250-300	—	250-600	250-400	—	375-425
5. Injection molding pressure, psi	—	8000-30000	8000-30000	10000-20000	—	—	8000-20000	8000-20000	—	15000-40000
6. Compression ratio	—	1.8-3.6	1.8-2.2	2.0	—	—	—	—	—	2.0-2.5
7. Mold (linear) shrinkage, in./in.	—	0.015-0.050	0.015-0.050	0.02-0.05	—	0.020-0.050	0.015-0.035	0.007-0.011	—	0.003-0.007
8. Specific gravity (density)	D792	0.910-0.925	0.926-0.940	0.941-0.965	0.95-1.45	0.93-1.40	0.93	0.935-0.950	0.94	1.49-1.58
9. Specific volume, in.³/lb.	D792	30.5-30.0	30.0-29.6	29.6-28.8	—	—	—	—	29.8	18.4-17.8
10. Refractive index, np	D542	1.51	1.52	1.54	—	—	—	—	—	—
Mechanical										
11. Tensile strength, psi	D638, D651	600-2300	1200-3500	3100-5500	1600-4600	1500-3100	800-2000	1440-2800	2500	7500-9000
12. Elongation, %	D638	90.0-800.0	50.0-600.0	20.0-1000.0	10.0-440.0	180.0-600.0	300.0-700.0	750.0-900.0	525.0	15.0-65.0
13. Tensile modulus, 10⁵ psi	D638	0.14-0.38	0.25-0.55	0.6-1.8	0.5-5.0	—	0.046-0.067	0.02-0.12	1.02	3.60-4.75
14. Compressive strength, psi	D695	—	4800-7000	2700-3600	2000-5500	—	—	—	—	9000-22000
15. Flexural yield strength, psi	D790	—	0.5- > 16.0	—	2000-6500	—	3000-3600	—	—	14500-17000
16. Impact strength, ft. lb/in. of notch ½ X ½ in. notched bar, Izod test	D256	No break	—	0.5-20.0	1.0-20.0	—	No break	No break	No break	1.0-7.0
17. Hardness, Rockwell	D785	D41-D46 (Shore), R10	D50-D60 (Shore), R15	D60-70 (Shore)	33-80 (Shore D)	33-57 (Shore D)	D27-36 (Shore)	D17-38 (Shore)	D60-70 (Shore)	R117-122
18. Flexural modulus, psi X 10⁵	D790	0.08-0.60	0.60-1.15	1.0-2.6	0.7-3.5	—	0.01-0.20	0.01-0.20	—	3.8-4.5
19. Compressive modulus, psi X 10⁵	D695	—	—	—	0.5-1.5	—	—	—	—	3.35-6.00
Thermal										
20. Thermal conductivity, 10⁻⁴ cal./sec/cm²/(°C/cm)	C117	8.0	8.0-10.0	11.0-12.4	—	—	—	—	—	3.3
21. Specific heat, cal./°C/g (RT)	—	0.55	0.55	0.55	—	—	0.55	0.55	—	0.33
22. Thermal expansion, 10⁻⁵/°C	D696	10.0-20.0	14.0-16.0	11.0-13.0	10.0-35.0	10.0-35.0	16.0-23.0	16.0-20.0	7.2	6.8-7.6
23. Resistance to heat, °F (continuous)	—	180-212	220-250	250	275	275	190-200	—	230	230
24. Deflection temp., °F @ 264 psi fiber stress	D648	90-105	105-120	110-130	100-175	100-175	—	93	163	202-234
@ 66 psi fiber stress		100-121	120-165	140-190	—	—	140-147	140-147	—	215-247

Table 2.2.15. (continued)

		ASTM								
Electrical	25. Volume resistivity, ohm/cm^3 (50% RH and 23°C)	D257	$> 10^{16}$	$> 10^{16}$	$> 10^{16}$	$< 200\text{-}{>}15 \times 10^8$	2.4×10^9	1.5×10^8	$> 10^{16}$	10^{15}
	26. Dielectric strength, short-time, 1/8-in. thickness, V/mil	D149	450-1000	450-1000	450-500	230-1420	450-550	620-780	710	1220-1500
	27. Dielectric strength, step-by-step, 1/8-in. thickness, V/mil	D149	420-700	500-700	440-600	—	—	620-780	680	—
	28. Dielectric constant, 60 cycle	D150	2.25-2.35	2.25-2.35	2.30-2.35	2.28-7.60	2.7-2.9	2.50-3.16	—	3.08
	29. Dielectric constant, 10^3 cycle	D150	2.25-2.35	2.25-2.35	2.30-2.35	2.27-7.40	2.7-2.9	2.60-2.98	—	2.8-3.6
	30. Dielectric constant, 10^6 cycle	D150	2.25-2.35	2.25-2.35	2.30-2.35	2.63-3.11	2.7-2.8	2.6-3.2	2.30	3.2-3.6
	31. Dissipation (power) factor, 60 cycle	D150	< 0.0005	< 0.0005	< 0.0005	0.003-0.044	0.01-0.02	0.003-0.020	—	0.01887-0.03080
	32. Dissipation (power) factor, 10^3 cycle	D150	< 0.0005	< 0.0005	< 0.0005	0.00048-0.04900	0.01-0.02	0.011-0.017	—	0.0092-0.0108
	33. Dissipation (power) factor, 10^6 cycle	D150	< 0.0005	< 0.0005	< 0.0005	0.001-0.002	0.01-0.02	0.03-0.05	0.0002	0.020
	34. Arc resistance, sec	D495	135-160		200-235					
Resistance characteristics	35. Water absorp., 24 hr, 1/8-in thick, %	D570	< 0.015	< 0.01	< 0.01	< 0.01-0.06	0.04	0.05-0.13	< 0.01	0.02-0.15
	36. Burning rate (flammability), in./min.	D635	Very slow (1.04)	Very slow (1.00-1.04)	Very slow (1.00-1.04)	Very slow to self-exting.	Very slow	Very slow	Very slow	Nonburning
	37. Effect of sunlight	—	Unprotected material crazes rapidly. Requires black for complete protection, but weather-resistant grades available in natural and colors.			Same as for Polyethylene	Very slight yellowing	Very slight yellowing	—	—
	38. Effect of weak acids	D543	Resistant	Very resistant	Very resistant	Very resistant	Resistant	Resistant	—	None
	39. Effect of strong acids	D543	Attacked by oxidizing acids	Attacked slowly by oxidizing acids	Attacked slowly by oxidizing acids	Attacked slowly by oxidizing acids	Attacked by oxidizing acids	Attacked	—	None
	40. Effect of weak alkalies	D543	Resistant	Very resistant	Very resistant	Very resistant	Resistant	Resistant	—	None
	41. Effect of strong alkalies	D543	Resistant	Very resistant	Very resistant	Very resistant	Resistant	Resistant	—	None
	42. Effect of organic solvents	D543	Resistant (below 60°C)	Resistant (below 60°C)	Resistant (below 80°C)	Resistant (below 80°C)	Attacked by chlorinated solvents; soluble in aromatic solvents above 50°C	Soluble in chlorinated and aromatic solvents over 50°C	—	Resists most
	43. Machining qualities	—	Good	Good	Excellent	Good to excellent	Fair	Fair	—	
	44. Clarity	—	Transparent to opaque	Transparent to opaque		Transparent to opaque	Translucent	Transp.-opaque	—	Excellent
	Transmittance, %	—	0-75	10-80	0-40	—	—	0-80	—	—
	Haze, %	—	4-50	2-40	10-50	—	—	2-40	—	—

From *Modern Plastics Encyclopedia 1969–1970*, McGraw-Hill, New York. With permission.

Table 2.2.15. (continued)

Table 2.2.15.
VARIOUS CHARACTERISTICS OF SELECTED RUBBERS AND PLASTICS

Part B

| Plastics properties chart | A.S.T.M. test method | Fluoroplastics | | | | Silicones | | | | Styrene-butadiene thermo-plastic elastomers |
		Polychloro-trifluoro-ethylene	Polytetrafluoro-ethylene molding compound	FEP fluoroplastic	Poly-vinylidene fluoride	Cast resins — Flexible (Including RTV)	Molding compounds — Asbestos filler	Molding compounds — Glass fiber filler	Molding compounds — Mineral filler	
Processing										
1. Molding qualities	—	Excellent	—	Excellent	Excellent	—	Good	Good	Excellent	Good
2. Compression molding temp., °F	—	460-550	—	600-750	400-550	—	300-350	310-360	310-375	250-325
3. Comp. molding pressure, psi	—	500-15000	2000-5000	1000-2000	500-1500	—	1000-8000	1000-5000	200-5000	100-3000
4. Injection molding temp. °F	—	500-600	—	625-760	450-550	—	—	—	—	300-425
5. Injection molding pressure, psi	—	20000-60000	—	5000-20000	15000-20000	—	—	—	—	15000-30000
6. Compression ratio	—	2.0	3.0-4.0	2.0	1.8 plts. 3.6 pdr.	—	6.0-8.0	6.0-9.0	1.7-2.0	2.0
7. Mold (linear) shrinkage, in./in.	—	0.010-0.015	—	0.03-0.06	0.030	—	—	0-0.005	0.003-0.006	0.001-0.005
8. Specific gravity (density)	D792	2.1-2.2	2.14-2.20	2.12-2.17	1.75-1.78	0.99-1.50	1.6-1.9	1.68-2.0	1.81-2.82	0.93-1.10
9. Specific volume, in.³/lb.	D792	13.2-12.7	12.9-12.5	13.0-12.8	15.7-15.6	—	17.3-14.4	16.5-13.8	—	37.4-27.5
10. Refractive index, np	D542	1.425	1.35	1.338	1.42	1.43	—	—	—	1.52-1.55
Mechanical										
11. Tensile strength, psi	D638, D651	4500-6000	2000-5000	2700-3100	5500-7400	350-1000	28000-35000	4000-6500	4000-6000	600-3000
12. Elongation, %	D638	80.0-250.0	200.0-400.0	250.0-330.0	100.0-300.0	100-300	25.0-30.0	—	—	300.0-1000.0
13. Tensile modulus, 10⁵ psi	D638	1.5-3.0	0.58	0.5	1.2	900.0	—	—	—	0.008-0.500
14. Compressive strength, psi	D695	4600-7400	1700	—	8680	100	—	10000-15000	13000-18000	Max. stress 5000 no break
15. Flexural yield strength, psi	D790	7400-9300	—	—	—	—	3000-35000	10000-14000	7000-8000	Max. stress 60-900 no break
16. Impact strength, ft. lb./in. of notch (½ × ½ in. notched bar, Izod test)	D256	2.5-2.7	3.0	No break	3.6-4.0	—	—	3.0-15.0	0.26-0.35	No break
17. Hardness, Rockwell	D785	R75-R95	D50-D65 (Shore)	R25	D80 (Shore)	15-65 (Shore A)	—	M84	M71-M95	40-90 (Shore A)
18. Flexural modulus, psi × 10⁵	D790	—	—	—	2.0	—	—	11.6	—	0.04-1.50
19. Compressive modulus, psi × 10⁵	D695	—	—	—	1.2	—	—	—	—	0.036-1.200
Thermal										
20. Thermal conductivity, 10⁻⁴ cal./sec/cm²/1°C/cm)	C117	4.7-5.3	6.0	6.0	3.0	3.5-7.5	—	7.51-7.54	11.0-13.0	3.6
21. Specific heat, cal./°C/g, (RT)	—	0.22	0.25	0.28	0.33	—	—	0.24-0.30	—	0.45-0.50
22. Thermal expansion, 10⁻⁵/°C	D696	4.5-7.0	10.0	8.3-10.5	8.5	25.0-30.0	—	0.8	2.0-4.0	13.0-13.7
23. Resistance to heat, °F (continuous)	—	350-390	550	400	300	500	—	>600	>600	130-150
24. Deflection temp., °F	D648									
@ 264 psi fiber stress	—	258	250	—	195	—	—	>900	>900	Sub zero-150
@ 66 psi fiber stress		—	—	—	300	—	—	>900	—	Sub zero-120

Table 2.2.15. (continued)

	Property	ASTM								
Electrical	25. Volume resistivity, ohm/cm³ (50% RH and 23°C)	D257	1.2×10^{18}	$> 10^{18}$	$> 2.0 \times 10^{18}$	2.0×10^{14}	2.0×10^{15}	$10^{10}\text{-}10^{14}$	10^{14}	5.0×10^{13} -2.5×10^{16}
	26. Dielectric strength, short-time, 1/8-in. thickness, V/mil	D149	500-600	480	500-600	260	550	200-400	200-400	420-520
	27. Dielectric strength, step-by-step, 1/8-in. thickness, V/mil	D149	450-550	430	—	—	550	124-300	380	420-540
	28. Dielectric constant, 60 cycle	D150	2.24-2.8	<2.1	2.1	8.4	2.75-4.20	3.3-5.2	3.5-3.6	2.5-3.4
	29. Dielectric constant, 10^3 cycle	D150	2.3-2.7	<2.1	2.1	7.72	—	3.2-5.0	—	2.5-3.4
	30. Dielectric constant, 10^6 cycle	D150	2.3-2.5	<2.1	2.1	6.43	2.6-2.7	3.2-4.7	3.4-6.3	2.5-3.4
	31. Dissipation (power) factor, 60 cycle	D150	0.0012	<0.0002	<0.0003	0.049	0.001-0.025	0.004-0.030	0.004-0.005	0.002-0.003
	32. Dissipation (power) factor, 10^3 cycle	D150	0.023-0.027	<0.0002	<0.0003	0.018	—	0.0035-0.0200	—	0.001-0.003
	33. Dissipation (power) factor, 10^6 cycle	D150	0.009-0.017	<0.0002	<0.0003	0.17	0.001-0.002	0.002-0.020	0.002-0.005	0.001-0.003
	34. Arc resistance, sec.	D495	>360	>200	>165	50-70	115-130	150-250	250-420	95
Resistance characteristics	35. Water absorp., 24 hr, 1/8-in. thick, %	D570	0.00	0.00	0.01	0.04	0.12 (7 days @ 77°F)	0.1-0.2	0.08-0.13	0.19-0.39
	36. Burning rate (flammability), in./min.	D635	None	None	None	Self-extinguishing	Self extinguishing	None to slow	None to slow	Slow
	37. Effect of sunlight	—	None	None	None	Slight bleaching on long exposure	None	None to slight	None to slight	Slight color change
	38. Effect of weak acids	D543	None	None	None	None	Little or none	None to slight	None	None
	39. Effect of strong acids	D543	None	None	None	Attacked by fuming sulfuric	Slight to severe	Slight	Slight	Attacked by oxidizing acids
	40. Effect of weak alkalies	D543	None	None	None	None	Little or none	None to slight	None to slight	None
	41. Effect of strong alkalies	D543	None	None	None	None	Moderate to severe	Slight to marked	Slight to marked	Slight
	42. Effect of organic solvents	D543	Halogenated compounds cause slight swelling	None	None	Resists most solvents	Attacked by some	Attacked by some	Attacked by some	Dissolves
	43. Machining qualities	—	Excellent	Excellent	Excellent	Excellent	—	Fair	Fair to good	Poor
	44. Clarity	—	Transl. to opaque	Opaque	Transp. to transl.	Transp. to transl.	Clear grades available	Opaque	Opaque	Transp. opaq.
	Transmittance, %	—	—	—	—	—	—	—	—	—
	Haze, %	—	—	—	—	—	—	—	—	—

From *Modern Plastics Encyclopedia 1969—1970*, McGraw-Hill, New York. With permission.

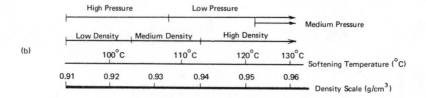

FIGURE 2.2.10.   Quantitative criteria for two scales for describing polyethylene characteristics.

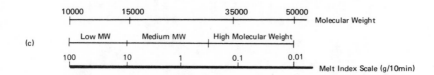

FIGURE 2.2.11.   Molecular structures of (A) butyl rubber and (B) ethylene-propylene copolymer.

Introducing a two-ply carbon/insulating paper named ''Duplex®'' successfully reduces the power factor. According to Reference 80, it has been established that higher impulse strength can be obtained by having the Duplex® layers as thin as compatible with mechanical requirements and with the insulating layer of high density and very high impermeability so as to provide a large barrier effect. Typical values are a total thickness of 0.1 mm (almost equally divided between the two plies), an impermeability of about 200×10^6 e.u., and a density slightly greater than unity. A conductor screen of this type was found to be excellent from both electrical and mechanical points of view. It gave optimum value of power factor and ensured the highest dielectric strength. The effect on power factor brought about by the insulation screen was found to be large when the screen stresses were high. Particular care should be taken therefore in the design of large conductor 33- and 66-kV cables. A new Duplex® design was proposed as shown in Figure 2.2.12.,[80] which confirmed experimentally that a considerable reduction in dissipation factor could be achieved.

B. Semiconducting Polymers Containing Carbon Blacks

Two types of semiconducting shields are being used in the U.S. in extruded-type cables;[84] one is a helically applied and lapped semiconducting tape and the second consists of an extruded semiconducting compound. Extruded strand and insulation shields are superior to tape shields in that they provide practically void-free contact with the insulation. Extruded-type shields made of a carbon black, dispersed in a dielectric have received widespread attention recently.

Table 2.2.16.
VARIOUS CHARACTERISTICS OF COOLANT MATERIALS

Material	Molecular weight	Content in air (wt%)	Tripple point (°K)	Tripple point (torr)	Latent heat for fusion (Kcal/kg)	Liquid specific weight (kg/ℓ)	Liquid equivalent to 1 Nm³ Gas (ℓ)	Specific heat (kcal/kgK)	Boiling point at 1 atm (K)
He	4.003	6.9×10^{-5}	—	—	0.835	0.125(bp)	1.432	1.066	4.216
Ne	20.18	1.3×10^{-8}	24.57	325	3.98	1.204(bp)	0.748	0.445	27.09
Ar	39.95	1.288	83.9	516.3	7.03	1.400_8(bp)	1.273	0.252	87.4
Kr	83.80	3.0×10^{-4}	104	548	4.67	2.155(126K)	1.735	0.128	121.3
Xe	131.3	3.9×10^{-5}	133	612.2	4.18	3.52 (171K)	1.664	0.172	164.0
H_2	2.016	3.5×10^{-6}	13.81	52.8	13.9	0.0708(bp)	1.270	2.231	20.27_3
N_2	28.01	75.520	63.15	94.0	6.1	0.812(bp)	1.539	0.492	77.34_9
O_2	3.200	23.142	54.34	1.2	3.31	1.144(bp)	1.252	0.404	90.18
Air	28.96	100	—	—	—	0.860(bp)	1.502	—	80.1
CO	28.01	0 ~ tr	68.0	115.38	7.18	0.791(bp)	1.580	0.532	81.64
NO	30.01	—	109.8	17.0	18.3	1.269(bp)	1.055	—	121.15
H_2O	18.02	—	273.15	4.58	79.76	1.000(281K)	0.8038	1.002	373.15
H_2S	34.08	—	187.7	13.78	16.69	0.914(bp)	1.664	0.478	212.8
SO_2	64.06	$(0 \sim 2) \times 10^{-4}$	197.7	12.56	27.5	1.46 (bp)	1.958	0.323	263.1
CO_2	44.01	4.8×10^{-2}	216.6	5.28 atm	4.32	1.178(217K)	1.696	0.433	—
N_2O	44.01	8×10^{-4}	170.8	659.17	35.33	1.225(bp)	1.603	—	183.6
CH_4	16.04	1×10^{-4}	90.5	88	13.97	0.42 (bp)	1.704	0.822	111.7
C_2H_6	30.07	—	101.2	0.01	22.70	0.547(bp)	2.453	0.577	184.8
C_2H_4	28.06	—	103.8	0.88	24.9	0.570(bp)	2.196	0.669	169.3
C_2H_2	26.04	—	191.7	129 atm	23.06	0.519(250K)	2.241	—	189.6
C_3H_8	44.10	—	83.0	7×10^{-6}	19.1	0.582(233K)	3.380	0.502	230.8
C_4H_{10}	58.12	—	134.9	—	19.2	0.601(bp)	4.316	0.546	272.7
C_6H_6	78.11	—	278.7	36.1	30.4	0.879(293K)	3.965	0.468	353.2
NH_3	17.03	0 ~ tr	195.5	45.6	79.4	0.678(bp)	1.121	1.068	239.8
$CFC\ell_3$ Fr11	137.37	—	162.2	602.5	12.0	1.532(273K)	4.001	—	296.8
$CH_2C\ell_2$ Fr12	120.91	—	118.3	—	8.2	1.473(bp)	3.663	—	243.1
H_g	200.59	—	234.28	760	2.5	13.546(293K)	—	0.03382	629.88

Material	Latent heat for evaporation (kcal/Nm³)	Critical points (K)	Critical points (atm)	Critical points (g/cm³)	Gas specific weight (kg/Nm³)	C_p(15°C) (kcal/Nm³ °C)	Specific weight ratio	Thermal conductivity (0°C) (kcal/m.hr.°C)	Viscosity (10^{-4} g/cm.sec)
He	0.875	5.20	2.26	0.0693	0.1786	0.2232	1.660	0.123	1.96
Ne	18.515	44.43	26.86	0.4835	0.9005	0.2232	1.64	0.0398	3.10
Ar	69.51	150.71	48.0	0.5308	1.7821	0.2232	1.667	0.0140	2.22
Kr	103.06	209.38	54.27	0.9085	3.739	0.2267	1.67	0.00763	2.46
Xe	134.74	289.74	58.0	1.105	5.858	0.2267		0.00446	2.26
H_2[a]	9.6	32.994	12.77	0.0308	0.08994	0.3070	1.410	0.150	0.88
N_2	59.6	126.14	33.49	0.3110	1.2499	0.310	1.406	0.0206	1.75
O_2[a]	72.8	154.78	50.14	0.430	1.42768	0.3134	1.396	0.0210	2.03
Air	63.42	132.5	37.2	—	1.292	0.3112	1.402	0.0208	1.81
CO	64.25	134.1	34.6	0.301	1.2497	0.312	1.404	0.0193	1.77
NO	146.78	180.2	64	0.52	1.3388	0.319		0.0200	
H_2O	434.2	644.05	218.3	0.323	0.8038	0.360	1.32	0.0208	8.80
H_2S	198.84	373.6	88.9	0.349	1.521	0.370		0.0110	
SO_2	265.9	430.3	77.7	0.524	2.8583	0.432		0.00702	(1.3)
CO_2	163.0	304.2	72.9	0.468	1.9635	0.392	1.31	0.0123	1.47
N_2O	176.5	309.6	71.7	0.457	1.964	0.394		0.0130	
CH_4	86.2	190.6	45.8	0.162	0.7158	0.377	1.31	0.0260	(1.14)
C_2H_6	155.48	305.5	47.2	0.203	1.3416	0.545	1.23	0.0157	
C_2H_4	144.06	282.8	50.5	0.227	1.2516	0.452	1.25	0.0145	
C_2H_2	230	309.2	61.6	0.231	1.1617	0.463		0.0158	
C_3H_8	200	368.7	44.0	0.220	1.9674	0.861		0.0130	
C_4H_{10}	238.7	425.2	37.5	0.228	2.5932	0.88		0.0120	
C_6H_6	328.5	562	48.6	0.300	3.4851	1.003		0.00754	
NH_3[a]	248.97	405.5	111.5	0.235	0.7599	0.394	1.32	0.0185	(1.03)
$CFCl_3$ Fr11	268.8	470.7	36.4	—	6.1293	—		—	
CF_2Cl_2 Fr12	215.25	384.2	39.6	—	5.3951	0.723			
H_g	69.45	>18.00	>2.00	4~5	8.9498	—	1.667	—	2.28

[a]   Explosion limit: NH_3, 13.5 to 79.0%; H_2, 4.65 to 93.9%; O_2, explosive with other inflammable materials.

Table 2.2.17.
PHYSICAL CONSTANTS OF COOLANTS AND OTHERS

Liquid	Temperature		Dielectric Constant	Density (g/cc)	Viscosity (micropoise)	Surface tension (dynes/cm)	Specific heat (cal/g/°C)	Heat of vaporiz. (cal/g)
	K	°C						
Helium	4.21	−269	1.0469^5	0.1251	31	0.12	1.08	6
	2.24		1.0563	0.1471	18	0.353 (2.5 K)	0.715	
	2.19		1.0563^5	0.1472	17.5	—	4.6^b	
	2.15		1.0565	0.1471	—^{a}	—	2.26	
	1.83		1.0562	0.1465	—^{a}	—	0.64	
Hydrogen	20.4	−253	1.231	0.0712	142	1.91	2.34	108
	14.0		1.259	0.0772	234 (15 K)	2.88	1.75	
Nitrogen	77.3	−196	1.431	0.881	1,600	8.27 (80 K)	0.48	47.6
	63.1		1.467	0.870	2,900	10.53 (70 K)	0.48	
Water	273	0	88	0.9999	17,980	75.6	0.9985	596
	293	20	80	0.998	10,020	72.75	1.0074	585
Transformer oil	273	0	2.22	0.900	450,000	30.5 (23°C)	0.425 (30°C)	—
	370	100	2.12	0.835	110,000	26.2	—	
Air	273	0		0.001293	171	—	0.24	
(Gas — 760 mmHg)	293	20	1.00059	0.001205	184	—	—	

Note: Calculated compressibility — cm²/dyne × 10^{12}.

	Adiabatic	Isothermal
Hydrogen at 20 K	977	1534
Nitrogen at 77 K	159	—
Nitrogen at 74 K	146	295

a   Super fluid.
b   Specific heat reaches an undetermined sharp maximum greater than 4.6.

Table 2.2.18.
AC AND DC BREAKDOWN STRENGTHS OF VARIOUS INSULATING MATERIALS AT CRYOGENIC TEMPERATURES

Material	Thickness (μm)	AC Breakdown Strength (peak value) (MV/cm)						DC Breakdown Strength (MV/cm)			
		Air (293°C)	LN_2 (77 K)	GN_2 (77 K)	LHe (4.2 K)	GHe (4.2 K)	Vacuum (10 K)	LN_2 (77 K)	GN_2 (77 K)	LHe (4.2 K)	GHe (4.2 K)
Mylar®	100	1.19	1.71	0.99	1.20	0.85	2.12	2.1	1.6	2.1	1.5
Teflon®	100	1.13	1.14	0.97	1.34	0.99	3.20	2.0	1.7	2.3	1.4
Teflon® FEP	100	1.27	1.37	1.12	1.44	1.145	2.54	2.1	1.7	2.2	1.6
Polyethylene	100	1.02	1.55	1.06	0.92	0.70	1.45	1.8	1.2	1.8	1.1
Nylon	100	1.10	1.27	0.99	1.11	0.73	2.0	2	1.8	2.4	1.6
PVC	100	0.12	1.09	0.97	1.05	0.75	1.82	1	—	2	1.5
Polyimide-Kapton®	100	1.40	1.98	1.27	1.63	1.27	3.17	2.4	1.8	2.6	1.9
Kraft® paper	100	0.13	0.28	0.18	0.38	0.21	—	0.6	0.3	0.54	0.4
Mica paper	100	0.64	0.76	0.62	0.55	0.30	—	1.1	0.95	0.85	0.7
Kapton + Teflon®	50 + 50	1.34	1.51	1.4	1.47	1.17	3.34	2.4	1.7	2.5	1.8
Teflon® PTFE impregnated glass cloth	100	0.32	0.34	0.30	0.58	0.4	—	0.50	0.36	0.45	0.3

Note: Sphere (30 mm φ)-plate electrode, 0.5 kV/sec.

Table 2.2.19.
SHORT-TIME BREAKDOWN VOLTAGES OF SOME SOLID DIELECTRICS AT CRYOGENIC TEMPERATURES

Material	Air (296 K)	Liquid nitrogen 77 K	Liquid helium 4 K
0.002-in. Teflon®	4,700 V	6,300 V	6,000 V
0.004-in. Teflon® (2 sheets 0.002 in.)	9,300	10,400	9,600
0.005-in. Teflon®	5,800	9,000	8,100
0.010-in. Teflon®	12,300	14,000	12,200
0.002-in. Condenser paper (varnished)	3,600	6,800	4,800
0.004-in. Condenser paper (varnished) (2 sheets 0.002 in.)	5,700	11,100	9,400
0.005-in. Kraft® (varnished)	6,500	9,400	7,700
0.010-in. Kraft® (varnished)	11,700	13,800	5,500
0.002-in. polyethylene	6,300	11,500	6,500
0.001-in. Mylar®	2,300	4,600	3,500

Table 2.2.20.
BREAKDOWN STRENGTHS OF VARIOUS MATERIALS AT LIQUID NITROGEN TEMPERATURE

Material	Thickness (mm)	AC strength kV	AC strength kV/mm	Impulse strength kV	Impulse strength kV/mm	Impulse ratio
HD Polyethylene (Staflen® E605-M)	0.1	15.9 $\sigma = 0.5$	160	23.0 $\sigma = 2.1$	230	1.44
Polycarbonate (Yupilon®)	0.1	11.6 $\sigma = 1.0$	116	17.0 $\sigma = 1.1$	170	1.55
PET (Lumilar®)	0.05	7.2 $\sigma = 0.3$	144	12.7 $\sigma = 1.03$	253	1.76
Teflon®	0.05	7.2 $\sigma = 0.3$	145	14.3 $\sigma = 0.8$	259	1.79
Polysulfon	0.14	15.2 $\sigma = 1.5$	105	21.7 $\sigma = 3.4$	192	1.83
Polypropylene	0.028	7.8	278	10.3 $\sigma = 0.8$	372	1.44
Nylon (biaxially oriented)	0.03	5.8 $\sigma = 0.2$	194	12.3 $\sigma = 2.0$	411	2.12
PPO	0.1	13.1 $\sigma = 4.1$	131	28.7 $\sigma = 8.5$	287	2.20
Kraft® paper	0.125	7.6 $\sigma = 0.3$	61	20.3 $\sigma = 2.7$	157	2.59
Normex®	0.065	7.7 $\sigma = 0.3$	119	11.0 $\sigma = 2.8$	169	1.42
Cotton cloth	0.26	7.5 $\sigma = 0$	29	10.3 $\sigma = 2$	38	1.31
Vinylon cloth (Papilon®)	0.13	7.4 $\sigma = 0.4$	57	10.0 $\sigma = 3.1$	83	1.46

Table 2.2.21.
DISSIPATION FACTORS AND DIELECTRIC CONSTANTS OF NONPOLAR MATERIALS AND FILLED EPOXY RESINS

Material	tan δ		ε	
	In air 23°C	In LHe 4.2 K	In air 23°C	In LHe 4.2 K
Tetrafluoroethylene	0.0003	0.00002	2.0	—
Polyethylene	0.0041	0.00017	3.15	—
Polycarbonate	0.0021	0.000095	3.04	—
Mica (for reference)	0.00124	0.00082	7.8	7.6
Filled epoxy resin				
TiO_2 filler	0.012	0.0031	13.3	10.4
$BaTiO_2$ filler	0.013	0.0059	9.9	7.0
$SrTiO_2$ filler	0.013	0.0040	15.1	14.0

Table 2.2.22.
DISSIPATION FACTORS (500 Hz)

tgδ	293 K	77 K	4 K
Mylar ®	4×10^{-3}	10^{-3}	$1,8 \times 10^{-4}$ 2×10^{-4a}
Teflon® FEP	10^{-4}	2×10^{-4}	10^{-3} 3×10^{-3a}
Teflon® PTFE	10^{-4}	5×10^{-3}	10^{-3}
Polyimide (Kapton ®)	3×10^{-3}	2×10^{-3}	3×10^{-4}

[a]   Caloric Method 50 Hz.

Table 2.2.23.
ELECTRICAL PROPERTIES OF POLYMER PAPERS

Polyethylene papers (Tyvek®)	Polypropylene paper	P30 paper
LN_2 impregnated tan δ ≃ 10^{-4} $ε^* \sim 2$ (PE film 2.3) BDV: cable model 35 kV/mm	Polybutene impregnated tan δ ≤ 7×10^{-4} $ε^* ≤ 2.1$ (60 Hz, 80°C, 500 V)	Paraffin impregnated tan δ = $1.2 \sim 4 \times 10^{-4}$ (20°C $\sim$ 80°C) $ε^* = 2.5$ (20°C) BDV (100 hr, without slit) 100 $\sim$ 120 kV/ mm Cable model 60 kV/mm

There exist many types of carbon black which have conductivities ranging from 10^3 to 10^{-6} mho/cm depending on the graphite content.[84] For carbon-filled polymers, it is common practice to select highly conducting carbon types such as acetylenic blacks. Although pure graphite is the most conductive carbon type, it does not lend itself to this application because it is not readily available in the finely divided form required for the production of these composites. Recently a special kind of furnace carbon black named Ketjen® black has been developed and used on a trial basis to make polymers semiconductive.

Table 2.2.24.

BREAKDOWN STRESSES OF SELECTED SPECIMENS IMMERSED IN LIQUID NITROGEN (V/MIL)

Material	50% RH 10 psig	50% RH 75 psig	"Wet" 75 psig	100% RH 75 psig
Wood insulating paper	950	1150	—	—
(0.010-in. thick sheet)	1000	1150	—	—
	950	1200	—	1300
Nomex®	1200	1350	800	—
(0.010-in. thick sheet)	1150	>1350	650	—
	1150	>1350	800	—
Tyvek®	750	1025	—	—
(0.006-in. thick sheet)	700	1025	1220	—
	725	1000	1110	—
Calendared Tyvek®	—	>1350	—	—
(~0.005-in. thick sheet)	—	>1350	—	—
	—	>1350	—	—

Note: Decrease in pressure occurred with high voltage maintained and specimens failed prematurely. Small specimens: radial thickness of insulation — 0.040 in., carbon black screen at outer electrodes — 0.005 in.

Table 2.2.25.

BREAKDOWN STRESSES OF SELECTED SPECIMENS IMMERSED IN LIQUID HYDROGEN (V/MIL)

Material	50%RH 10 psig	50%RH 75 psig	"Wet" 75 psig	100%RH 75 psig
Calendered Tyvek®	800	1350	1350	—
(0.005-in. thick sheet)	700	>1350	1350	—
Wood insulating paper	—	>1350	—	—
(0.010-in. thick sheet)	—	>1000	450	1300
Nomex®	—	1250	—	1350
(0.010-in. thick sheet)	—	>1350	—	—
	—	1000	—	—

Note: Small specimens: radial thickness of insulation — 0.040 in., carbon black screen at outer electrode — 0.003 in.

From Jefferies, M. J. and Mathes, K. N., *Insulation System for Cryogenic Cable,* IEEE PES Winter Meeting TP44-PWR, IEEE, Piscataway, N.J., 1969, 6. With permission.

Among the polymers are butyl, styrene-butadiene, and neoprene rubbers as a polymer matrix. Ethylene copolymers, such as ethylene-propylene, ethylene-ethyl acrylate, and ethylene-vinyl acetate copolymers, and thermoplastics such as polyethylene and cross-linked polyethylene have also found wide application recently.

When the two components are mixed, the number of contacts among carbon black particles will be a function of both the particular particle shape and the particle concentration. Electronic conduction in such a system is thought to be due to electrons tunneling between carbon particles dispersed in the polymer matrix.

Table 2.2.27. shows various values of electrical resistivity for carbon-dispersed polymers. A semiconducting shield should have good conformity or contact with the polyethylene or

Table 2.2.26.
CHARACTERISTICS OF CELLULOSE
PAPER IMMERSED IN LIQUID NITROGEN

Impregnant	Voltage stress at breakdown (V/mil)
Oil at 23°C	725
Boiling liquid H_2 at 20 K	950
Boiling liquid N_2 at 77 K	1000
Freezing liquid H_2 at 14 K	1025

Voltage stress (V/mil)	Dissipation factor	
	Initial value	Value at end of voltage step
750	0.00087	0.00052
800	0.0011	0.00065
850	0.0038	0.00125
900	0.0097	0.0021

cross-linked polyethylene as well as sufficient electrical conductivity. For this reason a compound with a lower carbon content is preferred. Table 2.2.27. shows that the new carbon black is worthy of investigation. The material is characterized by a large surface to volume ratio (1000 m^2/g) and fine particles (30 μm).

C. Metal Shield [20]

Annealed copper tape has been satisfactory for both nonoverlapped and overlapped helical turns of shielding and also for double helical turns. Tinned copper is widely used to reduce the contact resistance between overlapped shielding layers. Cylindrical metal shields, formed by drawing down or direct extrusion of an aluminum tube, or by the continuous welding of a formed tape, have found a limited use with plastic-insulated cables. Characteristics of annealed copper are already presented in Table 2.2.1.

2.2.4. Metal Sheath

Metal sheaths belong to protective coverings and are normally used for oil-impregnated paper-insulated cables. They operate both to protect their insulation from probable mechanical damage and ingress of moisture or water, and sustain the mechanical effects of internal oil pressure and the electrical effects of fault currents. It is also required to be functional as an electric shield, except in the case of three-core cables. Lead sheaths have been widely used for these functional purposes. Recently, however, aluminum sheaths have increased in popularity because of their excellent mechanical performance and light weight. Iron sheaths may be used for economic reasons.

A lead alloy rather than high purity lead is normally used for sheathing because the pure lead is comparatively weak mechanically. Table 2.2.28. shows the atomic contents of lead alloy used for cable sheath.[3] The requirements of extrusion processability and mechanical strength for sheaths mandate the selection of an aluminum of lower purity than conductor-use aluminum. An example of composition of aluminum for sheath use is shown in Table 2.2.29.[3] Recently, a new type of sheath has been introduced.[37] It consists of a longitudinally corrugated, thin stainless steel or copper tape with a welded seam. This satisfies various performance requirements for a metal sheath, and at the same time provides sufficient mechanical strength and shielding capability. Table 2.2.30. shows typical characteristics of lead alloy, aluminum, and iron as sheath materials.[3]

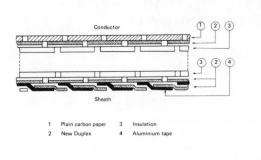

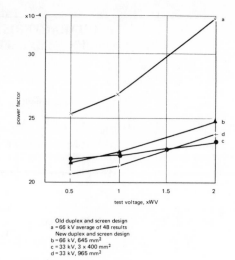

FIGURE 2.2.12.   Conductor and insulation screen with new duplex and dissipation factors. (From Miranda, F. J. and Gazzana Priaroggia, P., *Proc. Inst. Electr. Eng.*, 123(3), 230, 1976. With permission.)

The sheaths of OF cable are reinforced with stainless steel, brass, bronze, or annealed copper tapes to provide enough mechanical strength to counter the internal force of the pressurized oil. Stainless steel is one of the best reinforcing materials presently available because it is nonmagnetic, mechanically strong, reasonably noncorrosive, and not subjected to stress cracking.

Long-term mechanical performance with respect to such factors as fatigue and creep is also important in selecting a sheath metal. Fatigue is usually measured by the number of repeated cycles of weak stress or strain it takes to produce material fracture; this is referred to as the S-N characteristic.

Aluminum is superior to lead alloy in fatigue resistance, especially in the region of very weak strain. Creep is said to occur when the strain increases with time under a constant unidirectional stress. Creep in a metal sheath is due mainly to internal pressure which produces a circumferential (or tangential) force and a longitudinal force. The former, or hoop stress as it is called, is twice as great as the latter. Hoop stress is approximately 10 times the internal pressure, and therefore, usually 0.1 to 0.4 kg/mm^2; this compares with 2 kg/mm^2 for the tensile strength of the lead alloy sheath. Creep can be expected in a lead sheath when it is subjected to this hoop stress over a long period of time. For this reason, the lead sheath is reinforced with stainless steel tapes. No reinforcement is needed for aluminum and iron sheaths because they exhibit no creep when operated in the normal mode.

Corrosion is another problem for metal sheaths. Metallic corrosion takes place in the interface between the metal and its environment. It is accelerated by the presence of water and the degree of corrosion being specified by the electric potential induced when the metal is in contact with liquid of a certain pH. Many years of experience have shown that both lead and aluminum are virtually stable in normal ground water and sea water. It is, however, customary to apply a covering to protect the sheath against possible corrosive materials and the effect of stray electric currents from such sources as electric trains. This electrolytic corrosion due to stray current is precluded by insulated covering.

2.2.5. Protective Coverings

A. Cable Sheath and Anticorrosive Materials

Sheath and anticorrosive materials are required:

1. To have long-term stability as electrical insulation
2. To be resistant chemically to water and chemicals over a long period of time
3. To be antiweathering and antiaging
4. To have superior antiabrasion and antibending characteristics

Chloroprene, polyvinylchloride, and polyethylene are available for this purpose.

Chloroprene is a form of synthetic rubber which has superior mechanical properties and excellent chemically resistive and antiweathering properties. Since it has Cl-radicals in its molecular structure, as shown in Table 2.2.31.,[88] it is nonflammable. Its volume resistivity is of the order of 10^{12} Ω-cm, so it is much inferior as electrical insulation to natural rubber or styrene-butadiene rubber (SBR). Table 2.2.31. shows data for the electrical characteristics of a filled chloroprene. Cloth tapes may be lapped over chloroprene covering and subsequently vulcanized to produce an integral composite system. This reinforced system is mechanically tougher than any other two-material combination. However, it is difficult to apply on a cable with a rough surface, such as corrugated aluminum sheath. Also, it is rather expensive in material and processing, and damage can occur to the lead sheath and even the insulation as the temperature is raised for vulcanizing.

Polyvinylchloride, or PVC, is excellent regarding electrical properties and its resistance to chemical reaction. It is flame-retardant because of the Cl-radicals in its molecular structure. It can be endowed with various properties depending upon what plasticizers are introduced. It is characterized by fairly good processability and is relatively inexpensive. For these reasons, it is the primary material used for cable sheaths. PVC is classified into three categories, viz., soft, semihard, and hard ($< 5\%$) PVC, according to the plasticizer content. Semihard PVC is applied as a sheath material because of its mechanical strength and flexibility, although it is inferior in mechanical strength to chloroprene and polyethylene. It becomes brittle at low temperatures.

Of the three materials, polyethylene is the best as far as electrical properties are concerned; it shows excellent resistance to oil and chemicals and is good mechanically. On the other hand, it is less flame-retardant and less flame-extinguishing, for its molecular structure contains only carbon and hydrogen. Some polyethylene may exhibit environmental stress cracking.

Table 2.2.32. shows comparative data among the three materials, on which material selection can be based.[3]

B. Protective Materials

Various materials are utilized for the mechanical protection of cables. Lapped steel belts protect it from potential damage in installation and service operation. They are usually coated with anticorrosive paint. Steel wires, which may be applied for a submarine cable, vertical risers, bore hole, shaft and dredge cables can reinforce overall tensile strength of a cable. Steel wires are in wide use since they meet mechanical (50,000 to 70,000 psi) and anticorrosion requirements, but they do produce appreciable heat loss in the armor in the case of a single core cable. Nonmagnetic and highly conductive copper, aluminum, or aluminum alloy is preferable to reduce this kind of loss.

Among other cable protective materials are fibrous materials such as jute, jute cloth, cellulose paper, and cotton tape; filling materials such as asphalt and heavy oil, chemically resistant plastic films, and plastic pipes.

Polyethylene duct or pipe is widely used in the U.S. for installing extruded cable where direct burial is involved, the cable being pulled into the duct. It is referred to as combined duct cable or CD cable. The duct material is usually black polyethylene, with low or medium density.

Table 2.2.27.
RESISTIVITIES OF CARBON-FILLED POLYMERS

Polymer matrix	Kind of carbon	Carbon (wt %)	Resistivity (room temp.) (ohm-cm)	Remarks
XLPE	Acetylec C., Vulcan XC72	39	57 ± 3	DC measurement[84] Probe method[84]
XLPE	Acetylec C., Vukan XC72	32	744 ± 75	
XLPE	— [b]		35,200	60 Hz AC Bridge[83] Conductor shield use[83]
XLPE	— [b]		250	60 Hz AC bridge[83] Insulation shield use[83]
EPM	Acetylec C., Vulcan XC72	39	245 ± 25	DC measurement[84] Probe method[84]
EPM	Acetylec C., Vulcan XC72	32	17,700 ± 1,800	
—[a]	Ketjen® black	15	~100	Ref. 85
—[a]	Ketjen® black	15	~30	
—[a]	Acetylec black	40	~900	
—[a]	Acetylec black	40	~100	
XL-EVA	Acetylec black	40	~100	Ref. 86
XL-EEA	Acetylec black	14	~100	Ref. 86

[a] Rubbers and plastics in general.
[b] Probably acetylec carbon black.

2.2.6. Skid Wires and Steel Pipes
A. Skid Wires

Skid wires are approximately half oval in shape and made from hard copper, bronze, brass, or stainless steel wire (1.6 to 2.5 mm or 60 to 100 mil in radial height) with the flat side toward the shield. They are applied with a short lay to facilitate the pulling of the cables into the pipe. They should act to reduce the friction between their outer surface and the inner surface of the pipe, and be sufficiently strong mechanically to neither wear out nor fracture during installation. Stainless steel wire has recently received much attention for this function.

B. Steel Pipe

Oil-impregnated paper-insulated cables are installed in a carbon steel pipe which is then filled with oil under 200 psi or 14 kg/cm^2 pressure. The paper should be capable of withstanding a surge pressure as high as 1400 psi or 100 kg/cm^2 since such pressures may be generated at the time of a cable fault. Sections are usually 10 to 12 m or 30 to 40 ft long.

The pipe is internally coated with epoxy paint to prevent it from rusting. This antirusting paint must withstand lateral pressure and abrasion at the time the cable is pulled in. POF cable should not dissolve into the filling oil nor exert any harmful influence upon the electrical properties at operating temperature.

The pipe is covered externally with a somastic, tar epoxy, or polyethylene coating to retard corrosion. The coating, as a protective layer, should be characterized by:

1. Excellent and stable insulating performance
2. Excellent resistance to water and chemicals
3. Excellent weathering and aging properties
4. Good bonding to the pipe
5. Mechanical strength sufficient to withstand the weight of the steel pipe and such impact forces as attend installation
6. Continuity at steel pipe joints

Table 2.2.28.
ELEMENTARY CONTENT OF LEAD ALLOYS FOR CABLE SHEATH

Alloy	Standard	Sn	Sb	As	Cd	Bi	Cu	Te	Ag	Fe	Mg	Zn	Country
Pb	JIS H 2105	<0.015	<0.015	<0.010		<0.10	<0.010		<0.004	<0.010		<0.010	Japan
Pb	DIN 17640	<0.005	<0.005	<0.001		<0.005	0.03~0.05		< 0.001	<0.001	<0.001	<0.001	Belgium, France, Switzerland
Pb	ASTM B 29	<0.002	<0.002			<0.025	0.04~0.08		<0.002	<0.002		<0.001	U.S.
C	BS 801	0.4			0.15								Italy, France, Sweden
1/2C		0.2			0.075 ~ 0.08								Denmark, U.K., Italy, Spain, Holland, Canada, Sweden, France, Japan
E	BS 801	0.35 ~ 0.45	0.15 ~ 0.25										Japan, Italy, Germany, Sweden, U.K.
E'		0.4	0.12										
1/2E		0.2	0.1										
Te								0.04 ~ 0.06					Japan
Te + Cu	DIN 17640						0.02 ~ 0.05	>0.035					Germany, Japan
Cu	ASTM B 29						0.06						U.S.
E + Cu		0.4	0.2				>0.01						Germany
0.7Sb			0.7										Switzerland, U.K.
F-3		0.08 ~ 0.15		0.1 ~ 0.20		0.05 ~ 0.15							Norway, U.S., Canada
1/2F-3		0.06		0.075		0.05							
As + Sb	BNFMRA	0.2	0.2	0.015									

Note: JIS: Japanese Industrial Standard; DIN: Deutsches Institute für Normung; ASTM: American Society of Testing and Materials; BS: British Standard; BNFMRA: British Nonferrous Metal Research Association.

Table 2.2.29.
ELEMENTARY ANALYSIS OF
ALUMINUM FOR SHEATH USE

Si	Fe	Cu	Mg	Zn	Mn	Al
0.08	0.25	0.003	0.002	0.005	0.002	Rest

Table 2.2.30.
VARIOUS CHARACTERISTICS OF SHEATHING MATERIALS

	Lead alloy	Aluminum	Iron
Density (g/cm³)	11.3	2.70	7.86
Melting point (°C)	328	660	1530
Specific heat (cal, 20°C)	0.0309	0.22	0.11
Linear expansion coefficient (1/°C, 20°C)	29.1×10^{-6}	23.7×10^{-6}	11.7×10^{-6}
Thermal conductivity (cal/cm sec, °C)	0.0827	0.503	0.173
Tensile strength (kg/mm²)	$1.8 \sim 2.0$	8.5	33.0
Elongation (%)	45	33	42.5
Elastic modulus (kg/mm²)	1.8×10^3	7.2×10^3	18.0×10^3
Electrical resistivity (μ Ω-cm)	22	2.83	18
Conductivity (%)	7.8	60.9	9.5
Electrochemical equipment (mg)	1 .0737	0.0932	0.2894
Electrolytic quantity (g/Ah)	3.865	0.336	1.042
Electrolytic quantity (kg/A, year)	33.9	2.9	9.1

Table 2.2.33. shows various performance data for polyethylene lining.[3]

Carbon steel pipe has losses due to currents induced in it. Also, the AC resistance of its three conductors is increased by the increased magnetic flux density which arises from the high permeability of the carbon steel pipe. The pipe loss and the AC resistance of the conductors may be reduced by using stainless steel pipe.

2.3. ELECTRICAL BREAKDOWN CHARACTERISTICS

2.3.1. Electron Avalanche

The electron avalanche is a fundamental process that occurs in the electrical breakdown of all phases of matter — solids, liquids, and gases.[89-94] The original theory was first developed for gaseous breakdown by Townsend. Free or quasi-free electrons in a solid dielectric are accelerated by gaining energy from the external electric field and then collide with neutral atoms or molecules, or phonons (or modes of lattice vibration in crystalline solids) to create their progeny. The result is the cataclysmic transformation of an electrical insulator into an electrical conductor, i.e., electrical breakdown of the substance. This electron avalanche process, which describes the growth of an electron swarm from a single electron, is referred to as the single electron theory, because no energy distribution of electrons is taken into consideration.

Consider the simple picture of avalanche formation in which one electron, initially at the cathode, receives sufficient energy from the applied field to ionize a bound electron as shown in Figure 2.3.1. If both of these electrons now receive the same energy from the field, they will each cause a further ionization; eventually 2^n free electrons will be formed if there are n generations between cathode and anode. This is the so-called α mechanism for Townsend discharge. Such an avalanche may lead to breakdown if n is sufficiently large in the given circumstances, but generally speaking, it is not a critical condition for gaseous breakdown.

Table 2.2.31.
ELECTRICAL CHARACTERISTICS OF A FILLED CHLOROPRENE

Neoprene W	100	100	100
Magnesium oxide	4	4	4
Zalbe special	2	—	—
Akroflex CD	—	2	2
Hard clay	90	120	100
Whiting	—	—	45
Titanium oxide	5	—	—
FEF black	—	15	20
Light process oil	10	15	—
Kenflex A	—	—	30
Heliozone	—	4	4
Zinc oxide	5	5	5
NA-22 (vulcanized at 153°C for 10 min)	1	1	1

Electrical Properties (24°C) Specimen: 0.635 × 152 × 152 mm

Volume Resistivity (Ω-cm)	1×10^{12}	4×10^{11}	2×10^{13}
Permittivity (1 KHz)	6.0	7.5	7.0
Loss Tangent (1 KHz), (%)	2.0	3.3	3.5
Breakdown Strength (kV/mm)	30	25	24

Molecular Structure of Chloroprene

$$
\begin{array}{cccccccc}
& H & Cl & & H & Cl & & H \\
& | & | & & | & | & & | \\
-& C & -C & =C- & C & -C & =C- & C- \\
& | & & | & | & & | & | \\
& H & & H & H & & H & H
\end{array}
$$

The ionization coefficient, i.e., the number of electrons created in unit distance by ionization collisions can be written

$$\alpha(F) = (1/\lambda) \exp(-I/e\lambda F) \tag{2.3.1.}$$

where F = the applied electric field, I = the ionization potential, λ = the mean intercollision distance, or mean free path for ionization, and e = the electronic charge.

The 40 generation theory, given by Seitz,[95] is generally applied to solid dielectrics. One initial electron, starting at the cathode, grows into a swarm of electrons ($e^{\alpha d}$) at the anode through the mechanism of ionization by collision. Simultaneously, it diffuses laterally. A criterion for this breakdown is derived from the assumption that the energy given to a space of colliding electrons is equal to the total binding energy in the space. It is postulated that a 40-times collision makes a solid dielectric breakdown, i.e., $n = \alpha d = 40$. The critical field for breakdown can be derived as [93]

$$F_c \simeq H/1_n (d/F_c \lambda_n) \tag{2.3.2.}$$

where d = the interelectrode distance, λ = the mean free path, n = the number of collisions till breakdown, and H = the Hippel electric field which is related to the energy dependence

Table 2.2.32.
COMPARISON OF PERFORMANCES OF THREE MATERIALS

Item	Chloroprene	PVC compound	Polyethylene
Electrical characteristics			
Electric strength (kV/mm)	15 ~ 20	20 ~ 35	35 ~ 50
Volume resistivity (Ω-cm)	$10^7 \sim 10^{12}$	$10^{12} \sim 10^{15}$	10^{18}
Mechanical characteristics			
Tensile strength (kg/mm²)	0.8 ~ 2.0	1.0 ~ 2.5	1.2 ~ 1.5
Elongation (%)	300 ~ 1000	100 ~ 300	500 ~ 700
Abrasion resistance	Excellent	Excellent	Good
Weathering resistance	Good	Good	Good
Coldness resistance	Good (down to −40°C)	Excellent (down to −10°C)	Excellent (down to −40°C)
Flame resistance	Good	Good	Bad
Water resistance	Good	Good	Excellent
Oil Resistance			
Gasoline, heavy oil	Bad	Good	Excellent
Benzol	Bad	Bad	Good
Antichemicals			
Sulfuric acid	Bad	Good	Excellent
Sodium hydroxide	Good	Good	Excellent

Table 2.2.33.
VARIOUS PERFORMANCES OF POLYETHYLENE LINING

Item	Performance	Remarks
Tensile strength	150 ~ 200 kg/cm²	ASTM D 638-58T
Elongation	600 ~ 800%	ASTM D 638-58T
Hardness	Shore D 46 ~ 50	ASTM D 1706-61T
Softening point	Vicat 85 ~ 90°C	ASTM D 1725-58T
Volume resistivity	$>1 \times 10^7$ Ω·cm	
Ball drop test	17 ~ 38 times (2.4 mm thick)	Until pinhole generation
Thermal deformation test	100 psi; 2.33% 200 psi; 2.76%	
Stress cracking resistance	>100 hr	
Contactness	14 ~ 20 kg	180°C, 100 mm wide
Flatness test	To pass test at test temps	

of the collisions time and the probability that an electron makes no collision with the lattice during this time interval of the breakdown process. The approximate picture of the single electron avalance theory just described is believed to take place, in principle, but not exactly as represented by Equation 2.3.2. A many-electron theory (i.e., one taking explicit account of the distribution function of conduction electrons) of collision ionization breakdown has been developed.[93] In most gases and liquids, recombination is a factor; free electrons recombine with their departed ions within times as short as 10^{-9} to 10^{-6} sec after their generation. It is therefore preferable to use an effective ionization coefficient which includes attachment as well as ionization.

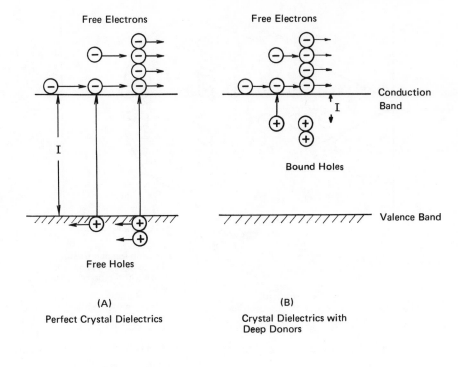

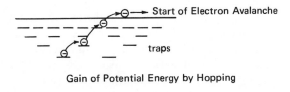

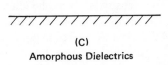

FIGURE 2.3.1.   Electronic states and avalanche processes in solid dielectrics.

In gases, other processes have been postulated as following the electron avalanche to finalize the breakdown as shown in Figure 2.3.2. The γ-mechanism is a positive feedback process in which positive ions, generated by ionization collisions, moved toward the cathode much more slowly than their companion electrons and collide with that electrode to produce secondary electrons. These in turn are accelerated by the electric field to produce further progeny. The breakdown criterion for this process is given by

$$\gamma(e^{(\alpha-\eta)d}-1) = 1 \qquad\qquad (2.3.3.)$$

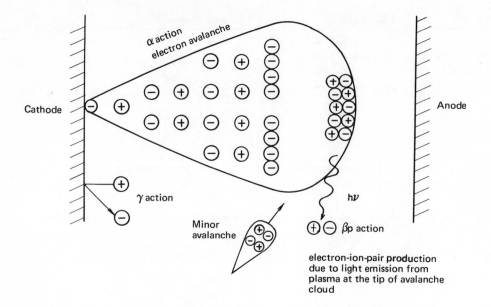

FIGURE 2.3.2.   Electron avalanche processes in gases.

where γ = the number of electrons emitted from the cathode per bombarding ion and by such photo ionization as may occur there, α = the ionization coefficient, and η = the attachment coefficient. The term $(\alpha - \eta)d$ should be replaced by the integral $\int^d_0 (\alpha - \eta)dx$ in the case of a nonuniform electric field, because $(\alpha - \eta)$ is a strong function of the electric field and therefore of the distance.

The streamer theory which involves γ action at the cathode and β_ρ action near the anode is another criterion. Photons emitted following collision in the gap produce free electrons (photo-ionization β_ρ) near the anode. These are again accelerated to produce avalanches. By these means, a streamer is formed which grows rapidly across the gap and leads to final breakdown. For long gaps, a leader or an ionization voltage wave is under investigation.

For liquids, it is generally accepted that electrons emitted from the cathode surface are accelerated by the field and retarded by collisions with liquid molecules. The mechanism whereby electrons are slowed down in liquids is different from that in gases.[94] In gases, an electron loses its energy mainly in ionization processes and in excitation collisions with atoms or molecules. In a liquid, since the density is several hundred times greater than that of a gas, the electrons are also acted upon by electrostatic forces (polarization) and are therefore influenced by their surroundings. The mean free path of an electron in a liquid is much smaller (about 10^{-7} cm), so it can acquire less energy than in a gas. Most of the energy lost by an electron appears as heat.

2.3.2. Intrinsic Breakdown

The term "intrinsic breakdown" is used mainly for solid dielectrics. An electron gains its energy from the electric field and transfers it to phonons. Dielectric breakdown occurs when the energy balance between gain and loss of energy is violated. This is the criterion for the intrinsic breakdown. It can be differentiated from avalanche breakdown by recognizing that the latter depends on the thickness of a dielectric, whereas the former does not.

The energy balance equation can be given by:

$$A(F, T_0, E) = B(T_0, E) \qquad\qquad (2.3.4.)$$

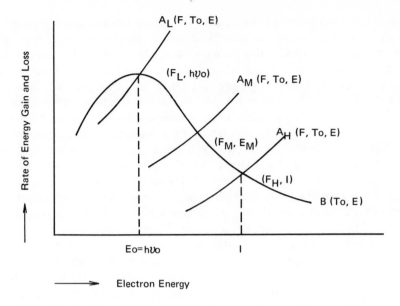

A_L (Fu, To, hνo)   : Low Energy Criterion
A_H (F_H, To, I)   : High Energy Criterion
A_M (F_M, To, E_M) : Intermediate Energy Criterion
                (Collective Breakdown Theories)

FIGURE 2.3.3.   Criteria for intrinsic breakdown.

where A = the rate of energy gain from the applied field, B = the rate of energy loss to the lattice or phonons, F = the applied field, T_0 = lattice temperature, and E = electron energy. It is assumed that the energy change per inelastic collision is much smaller than the electron energy. The energy gain A is given by

$$A(F, T_0, E) = e^2 F^2 \tau (E, T_0)/m \qquad (2.3.5.)$$

where $\tau(E,T_0)$ = the collision relaxation time for a conduction electron and m = the electron mass. The energy loss B is closely related to collision cross section and is dependent on the electron energy, having a peak at a certain energy as shown in Figure 2.3.3. It may be written as

$$B(E, T_0) = \sum_w h\nu(W) (P_w^e - P_w^a) \qquad (2.3.6.)$$

where hν(W) = a photon quantum with the azimuthal angle of W concerned with emission and absorption, P_w^e = the emission probability of a phonon W, and P_w^a = the absorption probability of a phonon w. Both quantities required for the energy balance (Equation 2.3.4.) can, in principle, be determined from these expressions.

According to the choice of values for the electron energy E, there are three theories or criteria available to explain intrinsic breakdown. They are

1.   The Fröhlich high energy criterion
2.   The von Hippel-Callen low energy criterion
3.   Collective breakdown theories

The first criterion assumes that the electron energy required for dielectric breakdown is the ionization energy I as shown in Figure 2.2.3. The critical value obtained from this consideration may be smaller than the true value because only electrons near the ionization energy fulfill the condition $A > B$. Criterion 2 claims that the breakdown takes place at an energy value where the rate of energy loss is a maximum, as shown in Figure 2.3.3. This corresponds to the emission and absorption of the optical phonon. It is clear that the evaluation of F_L amounts to finding a critical field that is able, on the average, to accelerate all conduction electrons against the retarding influence of the phonons. It has been objected that the critical field calculated from this consideration represents a criterion for breakdown which is too stringent, since other theories clearly lead to unstable situations at lower field strengths. Possibility 3 takes account of the energy distribution of the conduction electrons and gives an intermediate value for breakdown as shown in Figure 2.3.3. There are really two theories: the first applies to crystalline dielectrics with a negligible number of isolated electron energy levels and therefore with a low density of the conduction electrons (this is called the low density approximation and is essentially a type of avalanche theory); the second is the high density theory based on the assumption that the density of conduction electrons is such that they interchange energy with each other much more rapidly than with phonons.

The high density approximation may be applied to amorphous dielectrics, i.e., to a model in which a relatively large number of crystal imperfections gives rise to isolated electron energy levels below the conduction band as shown in Figure 2.3.1 (B). The following formula is pertinent:

$$F_c = (h\nu/\epsilon D\Delta V) \exp (\Delta V/2kT_0) \qquad (2.3.7.)$$

where $h\nu$ = the energy of phonon quantum, ϵ = the dielectric constant of the material of interest, D = constant, V = the energy depth of shallow traps, k = Boltzmann's constant, and T_0 = the lattice temperature. This is known as Fröhlich's theory for amorphous dielectrics. It assumes that free electrons interact with electrons trapped in shallow trap levels.

2.3.3. Thermal Breakdown

A solid dielectric fails thermally when the rate of energy or heat transfer to the lattice, due to an applied field F, exceeds the rate of heat absorption and heat dissipation. Such a condition leads to a catastrophic thermal runaway. The process is governed by the following equation:

$$C_\nu \frac{dT}{dt} - \text{div} (K \text{ grad } T) = \sigma F^2 \qquad (2.3.8.)$$

where C_ν = the specific heat per unit volume, K = the thermal conductivity, σ = the electrical conductivity, and T = the temperature, (dT/dt); the time derivative of the temperature. Equation 2.3.8. is independent of the type of electrical conduction, i.e., electronic or ionic conduction, and can therefore be applied universally.

A schematic diagram of solutions to Equation 2.3.8. is shown in Figure 2.3.4.; it assumes that a constant voltage is applied to the dielectric at time t = 0. Two extreme cases can be considered:

1.   Steady state thermal breakdown
2.   Impulse thermal breakdown

The first case assumes a quasi-steady state in the lattice as far as this affects the temperature of the hottest part of the dielectric. The thermal breakdown strength, F_m, can then be derived from the equation without the time dependent term

$$-\text{div} (k \text{ grad } T) = \sigma F^2 \qquad (2.3.9.)$$

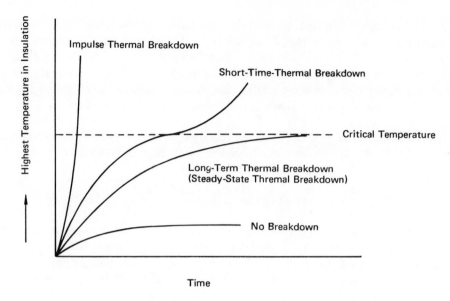

FIGURE 2.3.4. Schematic representation of temperature increase leading to final failure of insulation.

This gives the minimum thermal critical field appropriate to the situation where the field is applied for a very long time.

The second case occurs when the field is applied as a short pulse (of seconds or less in duration). It is assumed that a material fails before heat conduction starts. The critical field for this process can therefore be derived from the equation

$$C_V \, (dT/dt) \, = \, \sigma F^2 \qquad\qquad (2.3.10.)$$

Thermal breakdown is believed to take place at elevated temperatures, while electron avalanching occurs at low temperatures. Thermal breakdown can be classified with respect to the species of charge carrier and state as follows:

1. The rise of lattice temperature due to ionic conduction
2. The rise of lattice temperature due to the conduction of electrons in impurity levels
3. The rise of temperature of electrons in localized levels in amorphous substances

The first two processes are apparently similar and are based on the very positive dependence of electrical conductivity on temperature, as indicated by:

$$\sigma \, = \, \sigma_0 \exp \, (-U/kT) \qquad\qquad (2.3.11.)$$

which is sometimes called the Arrhenius relation. A solution may be obtained from the two formulas (2.3.8.) and (2.3.11.). A simple physical picture is as follows. Heat is generated in a dielectric due to Joule heating at a rate σF^2. Heat is absorbed in the dielectric at a rate of $C_v \, (dT/dt)$ and is transferred to the outside according to $-\text{div} \, (K \, \text{grad} \, T)$. Since the generated heat increases with rise of temperature, it may exceed the other two terms at a certain electric field, thereby leading to thermal runaway. This type of breakdown is characterized by a decrease in breakdown strength as temperature is raised. The phenomenon was first observed in Na^+ rich glasses. What is particularly important is that this process determines the limits of voltage and ampacity of EHV and UHV oil-impregnated cables, as will be described in the next chapter.

The third process corresponds to the Fröhlich's theory for amorphous dielectric which was treated in the previous section. Electronic thermal breakdown takes place within times as short as 10^{-9} to 10^{-8} sec, which is far less than the breakdown time for the usual thermal breakdown (above 10^{-3} sec), and it does not depend on thickness. This is because the specific heat of an electron system is much smaller than that of a lattice system. One might claim that the interaction of conduction electrons with phonons and the supply of energy to hopping electrons from the electric field are both neglected in the Fröhlich's theory. A new model has been proposed[96] to treat the rate of energy gain from the electric field of electrons hopping in the localized states and the rate of energy loss of the hopping electrons to phonons. In the simplest case the following formula applies:

$$F_B = C \exp(\alpha R) \exp(-\Delta V/2kT_0) /R \sqrt{\tau \nu_p} \qquad (2.3.12.)$$

where C = a constant, α = the tunneling coefficient, R = the interhopping distance, ΔV = thermal activation energy, T_0 = the lattice temperature, τ = the collision time, and ν_ρ = the attempt-to-escape frequency. It should be noted that the resultant critical field is similar to the Fröhlich's formula, but includes the interhopping distance explicitly.

2.3.4. Dielectric Breakdown in Polymers and Composite Materials

It is too difficult to apply any of the developed theories of breakdown, other than thermal breakdown which is macroscopic phenomenon, to polymeric and composite dielectrics. This is mainly because the material structure is much too complicated and breakdown is sensitive to so many factors. It might be affected, for example, by such materials characteristics as chemical structure, solid structure, molecular motion, internal defects (structural imperfections), and impurities (native foreign substances or additives).

Nevertheless, it is standard practice in this field of research to investigate the temperature dependence of breakdown in the hope that it may be possible to compare experimental results with some of the features of breakdown models which have been developed mainly for inorganic crystals. Figure 2.3.5. shows a general trend of temperature dependence of breakdown strength of polymers with and without polar radical groups.

The relations of dielectric breakdown with chemical structure may be summarized as follows:[70]

1. A critical temperature exists in nonpolar dielectrics, at which the breakdown strength abruptly changes. The strength is higher in the low-temperature region than in the high-temperature region. The glass transition temperature for a polymer is accepted as the separation temperature between the two.
2. The introduction of polar radicals into polymers increases breakdown strength.
3. Breakdown strengths of polymers are of the order of mega volt per centimeter at room temperature and are 2 to 10 times higher than that of ionic crystals.
4. Electron avalanching may take place in polymers at temperatures below the glass transition temperature. This assertion is substantiated by such experimental evidence as breakdown formative time lags,[97] thickness effects,[98,99] impurity effects, and so on. Dipolar vibrations derived from polar groups interact with accelerated electrons to reduce their speed; an increase of breakdown voltage is the result.
5. The high temperature region above the glass transition temperature can be separated into two subdivisions. In the highest temperature region, near the melting point, electromechanical breakdown is proposed. This is a deformation breakdown brought about by positive feedback of the Maxwell stress. A belief remains that in the final stage breakdown will take place electrically when the specimen thickness is reduced below a certain value because of the Maxwell stress (electrostrictive force).

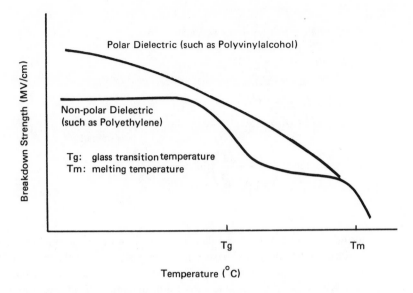

FIGURE 2.3.5.   General trend of temperature dependence of breakdown strength of polymers.

6.   In the high temperature region, between the glass transition temperature and the melting temperature, many models have been proposed to explain breakdown. These include pure thermal breakdown, electronic thermal breakdown, hopping electron thermal breakdown, and a quite different process — free volume breakdown.[100]

In amorphous polymers there are vacant spaces or holes between the molecules which are called free volume. Below the glass transition temperature, Tg, the free volume is approximately 2.5% of the total volume, but it increases with temperature above Tg. Artbauer,[100] in his theory, assumes that these randomly distributed holes can aggregate under electric stress. If the free volume is such that electrons are able to gain sufficient energy within it to cause breakdown, a breakdown criterion based on hole size can be derived. The length of the holes for this criterion is computed to be:

$$l_x = 0.5 \left\{ 1 - \frac{\log (0.217 \, N_e V p^2 t / \tau)}{\log [1 - (1 - p)^6]} \right\} d \qquad (2.3.13.)$$

where N_e = the density of electrons with energy sufficient to cause breakdown, V = the polymer volume, p = the specific free volume, t = the time during which the occurrence of l_x is considered, τ = the relaxation time of α molecular movement, and d = the hole diameter. A revised model has been proposed[101] which includes electron acceleration along the crystalline regions.

Oil-impregnated paper is a typical example of a composite material; it is one of the most excellent materials for cable insulation. The breakdown strength of oil-impregnated paper depends on the wave form of applied voltage, i.e., AC, DC, impulse, or switching surge voltages. This can be interpreted in terms of the distribution of voltage through this paper and oil. For example, the electric field is distributed capacitively for AC conditions and resistively for DC conditions. The impregnant oil has virtually zero voltage under DC voltage conditions. Breakdown of oil-impregnated paper is generally considered to start in the impregnant or with the ionization of a gas space in contact with an electrode. Experimentally obtained formulas for breakdown strength of oil-impregnated paper insulation are[102]

$$V = 50 + 50 \sqrt{\frac{\sqrt{A} \times \sqrt{d}}{\sqrt{t} \times 10^5}} \quad \text{(kV/mm) for AC} \qquad (2.3.14.)$$

$$V = 200 \sqrt{\frac{\sqrt{A} \times \sqrt{d}}{\sqrt{t} \times 10^5}} \quad \text{(thin paper)}$$

$$V = 100 + 200 \times \left(\frac{\sqrt{A} \times \sqrt{d}}{\sqrt{t} \times 10^5}\right) \text{(thick paper)}$$

$$\left.\begin{array}{c} \\ \\ \\ \end{array}\right\} \text{for impulse}$$

$$(2.3.15.)$$

where A = the airtightness (e.u.), d = the density (g/cm³) and t = the thickness (mm).

Even a solid dielectric such as polyethylene can be a composite material because it contains voids and macroscopic impurities. A thick specimen of the material gives a breakdown strength much lower than the "intrinsic value" for a thin specimen. Voids are usually filled with some kind of gas such as air or decomposition gases, which are vulnerable to ionization or corona discharge. Such a discharge can initiate solid breakdown. There is no direct evidence for this hypothetical process, but impregnating voids with a gas of high electrical strength or some additives increases the electric strength.

Most voids existing in extruded cable insulation are nearly spherical or ellipsoidal in shape. When a void is assumed to be spherical, the voltage across the void is given by:

$$V_q = E_0 d \; \frac{3\epsilon_1^*}{2\epsilon_1^* + \epsilon_2^*} \qquad (2.3.16.)$$

where E_0 = the electric field in the insulation, d = the diameter of a void, ϵ_1^* and ϵ_2^* = the specific dielectric constants of the extruded insulation and the void, respectively. The value of ϵ_2^* is essentially unity when the void is filled with gas. The electric field strength needed to initiate a gaseous discharge in a void can be calculated, assuming that the discharge follows the Paschen's curve even for a gap as short as 100 μm. The results of a computation for a void in which the pressure is 760 mmHg, and putting $\epsilon_1^* = 2\cdot3$, $\epsilon_2^* = 1$, and d = 5 μm ~ 100 μm are shown in Figure 2.3.6. This indicates that discharge initiated breakdown is probable in a dielectric with microvoids.

2.3.5. Liquid Dielectrics[88,94,103]

Several models have been proposed as mechanisms for dielectric breakdown in insulating liquids. They will be described briefly in the following.

A. Field Emission and Ionization by Collision

It is accepted that electrons emitted from the cathode surface are accelerated by the field and retarded by collision with liquid molecules. One electron produces $e^{(\alpha-\nu)d}$ positive ions near the cathode after it has traveled an interelectrode distance d. The positive ions so produced cause electron emission from the cathode. This positive feedback process leads to a catastrophic increase in electron population, thereby causing a breakdown of the liquid. This is similar to the Townsend discharge in gases. A big difference may be the loss mechanism. Since the density of liquids is several hundred times greater than gases, electrons collide with atoms or molecules much more often in liquids than in gases; furthermore, they are probably retarded by the polarization field which they themselves induce. The mean free

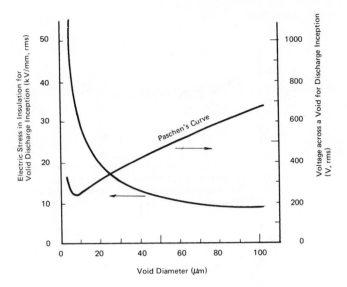

FIGURE 2.3.6.   Electric stress in insulation for void discharge inception as a function of diameter of a spherical void.

path of an electron is therefore much shorter (about 10^{-7} cm). The energy lost by an electron is transformed into atomic vibrations along chemical bonds, rotation of individual groups of atoms, and vibrations of the whole molecular chains, but most of it is changed into heat. (See Figure 2.3.7.[b].) This theory can explain the dependence of breakdown strength on voltage pulse width and gap length and also the distortion of the electric field in the liquid, but it does not explain the dependence of breakdown strength on pressure.

B. Bubble Theory[104]

The bubble theory was proposed to explain why breakdown of a liquid depends on pressure, while no electron multiplication process is observed during the prebreakdown stage. It is essentially a thermal breakdown theory. Electrons are injected into a liquid from an asperity on the cathode through field emission, as shown in Figure 2.3.7.(a), and give thermal energy to the liquid. The liquid breaks down when the energy is high enough to evaporate the liquid. The criterion for this breakdown is therefore:

$$AF^n \tau_r = m[C_p (T_b - T_0) + \ell_b] \qquad (2.3.17.)$$

where A = a constant, F = the local electric field strength, τ_r = the time for which a liquid sample remains in the region of highest stress, m = the quantity of liquid evaporated, C_p = the mean specific heat of the liquid from its specified temperature T_0 to its boiling point T_b, ℓ_b = the latent heat of evaporation, and n = the constant associated with the space charge-limited current from the cathode, $3/2 \sim 2$.

It is clear from Equation 2.3.17. that breakdown is dependent on pressure and material through T_b. Due to the high concentration of ions ($\sim 10^7$ cm^{-3}) at the cathode, the liquid is liable to overheat, and thereby causes the formation of vapor bubbles in the liquid. Electrons may be accelerated inside these bubbles and cause a gaseous discharge which may determine the process of breakdown of the liquid, as shown in Figure 2.3.7.(c). This leads to the following critical formula:

$$F_b^n = a^{-1} [C(T_b - T) + \theta] \qquad (2.3.18.)$$

where a, c, θ, and n are constants.

(a)   Ionic Current

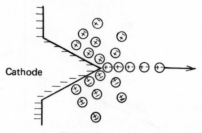

Cathode

Electrons are emitted from an asperity to attach
impurities. Resulting ions migrate to cause
frictional heat leading to the evaporation of the
liquid

(b)   Joule Heating

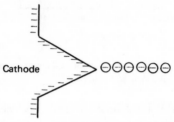

Cathode

Electrons are emitted from an asperity to cause
the Joule heating, leading to the evaporation of
the liquid

(c)   Ionization inside a Bubble

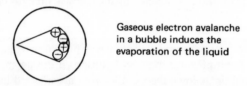

Gaseous electron avalanche
in a bubble induces the
evaporation of the liquid

(d)   Impurity

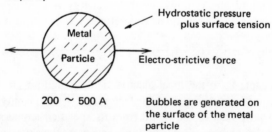

Hydrostatic pressure
plus surface tension

Electro-strictive force

Bubbles are generated on
the surface of the metal
particle

FIGURE 2.3.7.   Explanation of some bubble breakdown theories.

Another criterion can be derived on the assumption that breakdown takes place when the
electrostatic forces exceed the surface tension of bubbles,[105] as shown in Figure 2.3.7.(d).

C. Theory of Breakdown Due to Solid Particles

Polar molecules or macroscopic impurities tend to line up under the action of the electric
field, due to electrostatic forces, finally creating a bridge between the two electrodes which
causes breakdown.[106] The way in which impurity particles are pulled by the dielectrophoretic
force and balanced by the thermal diffusion force can be described as follows:

(a) Impurity

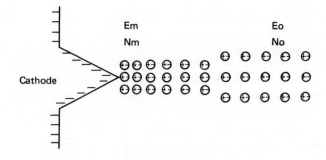

(b) Positive Ion Cluster (Iceberg Model)

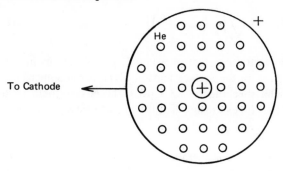

(c) Negative Ion Cluster

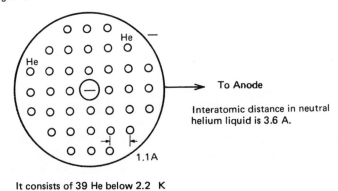

It consists of 39 He below 2.2 K
65 He at 4 K

FIGURE 2.3.8.   Explanation of some particle bridge theories.

$$kT\ell n \frac{N_m}{N_0} = 2\pi\epsilon_1^* \epsilon_0 \frac{\epsilon_2^* - \epsilon_1^*}{\epsilon_2^* + 2\epsilon_1^*} r^3 (E_m^2 - E_0^2) \qquad (2.3.19.)$$

where k = the Boltzmann's constant, T = the absolute temperature, ϵ_0 = the permittivity of vacuum, ϵ_1^* = the specific dielectric constant of the liquid, ϵ_2^* = the specific dielectric constant of the impurity material, r = the radius of an impurity particle, N_0 = the concentration of impurities near the electrode, E_0 = the average electric field, and E_m = the maximum electric field. (See Figure 2.3.8.[a]). Consideration of the viscosity of the liquid as counterforce against the dielectrophoretic force would lead to the following equation:

$$t_b g^4 r^7 (F_b^2 - F_0^2)^2 N^2 = A\eta^2 \tag{2.3.20.}$$

where N = the concentration of impurities, F_0 = the breakdown stress (DC breakdown), t_b = the time required for breakdown to occur (impulse duration), $F_b^2 \propto 1/t_b$, g = the coefficient for irregularities on the electrode surfaces, η = the coefficient of viscosity, and A = a constant. The results of this theory do not accord with experimental results as far as the absolute values for breakdown are concerned, nor with the observed dependence of breakdown strength on temperature and the diameter of impurities. However, it is valuable in explaining why breakdown strengths obtained for practical-use dielectric oils are usually much lower than the so-called "intrinsic" breakdown strengths.

There is another criterion for balancing the electrostrictive force on a conducting particle by hydrostatic pressure and surface tension. It gives[105]

$$F_b = 358 \sqrt{\left\{ \frac{1}{\epsilon} \left(\rho_\infty + \frac{2\sigma}{r} \right) \right\}} \tag{2.3.21.}$$

where ϵ = the dielectric constant of a liquid, ρ_∞ = the hydrostatic pressure (dyne per square centimeter), σ = the surface tension (dyne per centimeter), and r = the radius of an impurity particle. (See Figure 2.3.7.[d].) The electric field on the particle surface is considered to be four times greater than the average field.

D. Breakdown of Oils in Practical Use

Apart from theoretical consideration, the breakdown strength of a liquid dielectric in practical use is greatly influenced by the existence of impurities; it decreases drastically when water and fibers are present simultaneously.

2.3.6. Gaseous Dielectrics

Here we discuss the breakdown behavior of electronegative gases only, since they are the gases used for compressed-gas insulated cables.[103]

A. Breakdown Under Nearly Uniform Field

As described previously, an electron multiplication factor, represented by $\exp(\alpha - \eta)x$ is operative in electronegative gases when an electron travels a distance x. Terms α, η and $(\alpha - \eta)$ are called the first Townsend coefficient of ionization, the attachment coefficient, and the effective coefficient of ionization, respectively. When (E/p) is reduced to the point at which $\alpha = \eta$, no discharge takes place.[107] The discharge inception voltage for the conditions occurs when pd is comparatively large.

A linear relation between $(\alpha - \eta)/p$ and E/p is experimentally observed over the range of (E/p) for which discharges take place in SF_6 gas. It can be expressed:[108]

$$(\alpha - \eta)/p = 0.0277 \, E/p - 3.26 \tag{2.3.22.}$$

where $[(\alpha - \eta)/p] = [1/\text{cm} \cdot \text{torr}]$ and $[E/p] = [V/\text{cm} \cdot \text{torr}]$. The value of $(E/p)_{p \to \infty}$ is therefore 117.7. This linear relation justifies representing flashover voltage in the Paschen's law form. The following relation is obtained:[109,110]

$$V_s = 0.376 + 0.1179 \, pd \tag{2.3.23.}$$

where $[V_s] = [kV]$ and $[pd] = [\text{torr} \cdot \text{cm}]$. It is valid no further than atmospheric pressure. Further, it should be noted that when the pressure is low, breakdown strength falls below $(E/p)_{p \to \infty}$ in the region of large pd, due to electrode edge discharges.

For weakly nonuniform electric fields, it is widely accepted that discharge inception is governed by

$$\int_{\chi_1}^{\chi_2} (\alpha - \eta) \, dx = K \tag{2.3.24.}$$

where χ_1 = the distance at which the electric field is maximum, χ_2 = the dstance at which α equals η, and K = a constant (18 for negative impulse flashover voltage). Breakdown in SF_6 is characterized by flashover when the maximum electric field in a gap exceeds the limiting electric field by even a small increment, because $(\alpha - \eta)$ for SF_6 increases significantly with a small increase in the electric field. This may be called "maximum electric field dependence".

B. Particle-Initiated Breakdown

Below 1 $\sim$ 4 atm, discharge inception in SF_6 is determined by the effective coefficient of ionization. Above this pressure, other phenomena are observed: the deviation from Paschen's law, large scatter in flashover voltage, a significant conditioning effect, and an area effect. They are thought to occur as a consequence of ionic space charge, electrode surface effects, and floating dust.

Fine metal particles, as floating dust, cause a significant reduction in breakdown voltage of compressed SF_6 gas, especially in coaxial electrode systems, such as are used in compressed gas insulated (CGI) cables.[111,112] Free metal particles oscillate to and fro across the interelectrode gap. This so-called firefly phenomenon can cause breakdown. When metal particles are present and are in contact with an electrode, they acquire a charge whose magnitude depends upon the applied field and particle size. At a particular field, the particles will lift off and move towards the opposite electrode. Other mechanisms which might explain particle-initiated breakdown are[111]

1. The field at the particle surface may be sufficient to initiate breakdown, as it can for a fixed point electrode, before the lift-off occurs.
2. The particle may be lifted and cross the gap to touch the other electrode and initiate breakdown, as it might for a fixed point at that electrode.
3. After the charged particle has been levitated, but just before it hits the oppositely charged electrode, a micro-discharge occurs, which may be sufficient to trigger breakdown. This insulation instability must be suppressed for successful development of CGI cables. There are methods available to do this, such as particle traps and dielectric coatings of the conductor. These will be described in a later chapter.

C. Breakdown in Highly Nonuniform Fields

Discharge voltages under nonuniform field conditions can be divided into two groups: flashover voltage and corona inception voltage. It is characteristic of breakdown in compressed electronegative gases under nonuniform field conditions that the flashover voltage increases as pressure increases up to a certain value but then decrease as the pressure is increased still further. Such a peaking phenomenon is not observed in normal gases. The turn down in the flashover voltage of electronegative gases above a certain pressure is considered to be closely related to the formation of corona space charge prior to flashover. In other words, corona discharge suppresses flashover under low pressure, but leads to immediate flashover under high pressure.

D. Surface Discharge in Compressed Gases

Solid dielectric spacers are used to support conductors in the pipes of CGI cables. Higher electric stress is to be expected along a junction line where three kinds of materials meet (conductor, spacer, and compressed gas) than where no spacer exists. The result is a reduction in flashover voltage of the system. Since this flashover takes place along the surface of a spacer, it is referred to as a surface flashover. The ratio of gas flashover voltage to surface flashover voltage is called the spacer efficiency. Clearly, the spacer efficiency is determined by the probability of a corona occurring along the triple junction. This is greatly dependent on the dielectric constant of the spacer, the pressure of a compressed electronegative gas, and the micro-gap between the conductor and the spacer.

It is now a standard practice to extend the electrode into the spacer to shield the neighborhood of the triple junction electrostatically, and reduce the intensity of the electric field at the junction. It is possible to achieve a reduction of field along the surface of a disc-type spacer by establishing an angle between the surface of the spacer and the direction of the electric lines of force. There is a certain optimum angle for this purpose.

2.3.7. Dielectric Breakdown at Cryogenic Temperatures
A. Gaseous Breakdown

What has been learned in general about gaseous breakdown at cryogenic temperatures may be summarized as follows:[113]

1.   Paschen's law is valid for both DC and AC breakdown voltages of air and nitrogen gas, even at temperatures near their respective liquefying temperature, provided δd is less than 15, where δ is the relative gas density with respect to the density at 20°C and 1 atm, and d is the gap length in millimeters
2.   Paschen's law is valid in helium gas for δd up to 50 or 60
3.   No definite conclusion can be drawn for hydrogen gas because data is scarce and seemingly contradictory
4.   For air, nitrogen, and hydrogen, impulse breakdown voltage tends to increase at low temperatures in uniform electric fields
5.   Breakdown under nonuniform electric field conditions may be independent of irradiation at low temperatures.

As stated above, the breakdown mechanism in cryogenic gases is not very different from that of gases at room temperature. As we discuss breakdown phenomena in He, H_2, and N_2 gases, it will be noted that the fundamental difference lies in the breakdown mechanism of a monoatomic gas such as He and a molecular gas such as N_2.

Ionization occurs when energy is transferred to the atoms and/or molecules of a gas by electrons which have been accelerated by the external electric field. There is almost no energy transfer if the collisions are elastic. Inelastic collision must be involved in the excitation or ionization of atoms and molecules, and these require comparatively high energy gain electrons. This can be achieved in He by a sequence of several elastic collisions. For this reason ionization occurs in rare gases for a small value of Fλ (F = the electric field, λ = the mean free path) in spite of the high ionization energy of such gases. In a molecular gas such as N_2 or H_2, the low energy obtained by electrons in traveling their mean free path will be lost in exciting the atoms of molecules. In other words, the energy required for ionization should be obtained in a single free path, resulting in a large value of Fλ. It is therefore easy to understand why breakdown voltage is lower for He than for N_2 and H_2, and why it tends to increase as gas density increases.

B. Cryogenic Liquids

Possible mechanisms of electrical breakdown of cryogenic liquid dielectrics may be classified in the manner shown in Table 2.3.1.

Table 2.3.1.
MECHANISMS OF
CRYTOGENIC LIQUID
BREAKDOWN

Ionization by collision

Bubble model
    Ionic current
    Joule heating
    Ionization in a bubble
    Impurity

Particle-bridge model
    Impurity
    Positive ion cluster
    Negative ion cluster

Electron bubble model

a. Ionization by Collision

This is based on the belief that a process similar to collision ionization in gases takes place in cryogenic liquids. That this has not been confirmed experimentally appears to be a consequence of secondary effects.

b. Bubble Model

If a gaseous bubble exists for some reason, it is liable to cause ionization because of its low breakdown strength. A cryogenic liquid may be vaporized locally by the friction heat of ionic movement. Impurities in liquid He are estimated to number 10^{14} cm^{-3} for a purity of 99.99%, and to form micro-solidified gas particles at cryogenic temperatures. Particles with dipole moments tend to precipitate in high electric stress regions around electrode irregularities. When electron emission from the cathode begins, the emitted electrons collide with and attach to the particles. The resulting negatively charged particles leave the cathode and migrate toward the opposite electrode. Heat may be generated by friction between the moving charged particles and the ambient liquid, thereby inducing the formation of gas bubbles and finally leading to breakdown of the liquid. This is Gerhold's explanation of breakdown in liquid He.

Possible mechanisms due to joule heating, ionization within a gas bubble, and impurities are described in Section 2.3.5. The last effect, called Krauscki's theory, is classified as particle-initiation breakdown. Meats modified Krauscki's Equation (2.3.21.) and proposed the following two formulas:[115]

$$F_b = \frac{0.20}{\sqrt{\epsilon_0}} \sqrt{\left(P + \frac{2\sigma}{5 \times 10^{-9}}\right)} \text{ V/m} \qquad (2.3.25.)$$

for liquid He below the critical pressure

$$F_b = \frac{0.1}{\sqrt{\epsilon_0}} \sqrt{\left(P + \frac{2\sigma}{5 \times 10^{-9}}\right)} + 1.5 \times 10^7 \text{ V/m} \qquad (2.3.26.)$$

for liquid He above the critical pressure, where ϵ_0 = the permittivity of free space, farad/m, P = the applied pressure, N/m^2, and σ = the surface tension, N/m. While it is true that the results satisfy a breakdown criterion which closely resembles that proposed by Krauscki, it is to be noted that the breakdown mechanism is basically different; electron bubbles may be involved in the breakdown of liquid He, as will be described below.

c. Particle Bridge Model

Breakdown processes due to solid particle aggregation have been already described in Section 2.3.5. (See Figure 2.3.8[a].) Unlike impurity particles, positively charged solid clusters may be responsible for liquid helium breakdown. They consist of a group of polarized helium atoms around a positive ion. This is called an "iceberg",[116] a reverse concept to an electron bubble,[117] in which electrons or oxygen atoms are surrounded by a group of polarized helium atoms as if they were trapped in a solid cage structure as shown in Figure 2.3.8(b). Electrons confined in this structure are similar to polarons. According to Atkins,[118] the number of constituent helium atoms is 39 below the λ — point, and increases with temperature above the λ — point to become 65 at 4 K. The inter-atomic distance is 1.1 Å inside the cage, while it is 3.6 Å under normal conditions. This model can explain the polarity effect in a rod-plane electrode system and the gap length dependence of breakdown strengths.

d. Electron Bubble Model

Unlike other compressed gases, helium exhibits a marked change in behavior at very high densities, the breakdown field rising rapidly as the density is increased beyond 100 kg/m³.[115] There is a slight temperature dependence of the breakdown field for a given density in this region. As the density of gaseous helium is increased and becomes comparable with that of its liquid phase, electron mobility is dramatically reduced by three or four orders of magnitude.[119,120] This effect is ascribed to the local repulsion of helium atoms by an electron to form a pseudo-bubble or local region or rarified gas of 10^{-9}m radius, as shown in Figure 2.3.9. This concept is explained by Fowler and Dexter[117] as follows:

1. Liquid He has a negative electron affinity (i.e., a barrier) of about 1 eV.
2. The photo-ionization spectrum, which is sensible to bubble size, indicates a radius of about 20 Å.

Theoretical ideas which make this conclusion acceptable are based on (1) the Pauli principle, which gives rise to a strong He-electron repulsion, and (2) the low polarizability of He, which results in only a very weak attraction.

Accordingly, the electron is virtually trapped or localized and suffers a large diminution in mobility. The radius of these pseudo-bubbles is typically 2×10^{-9}m. Application of Gauss' law to an electron bubble of radius 5×10^{-9}m gives the field at the periphery as[115]

$$E = \frac{q}{4\pi\epsilon_0 R^2} = 57.5 \text{ MV/m} \qquad (2.3.27.)$$

which is approximately of the same magnitude as the observed breakdown fields. The application of this electric field would collapse a pseudo-bubble to release the trapped electron, prohibit further trapping, and accelerate the electron. Finally, it would lead to an electron avalanche and subsequent breakdown. Figure 2.3.10. shows helium breakdown for 1 mm gap as a function of helium density and pressure.[115]

C. Liquid Nitrogen

Breakdown strengths under DC voltage application obtained for liquid nitrogen are in the range 100 to 200 kV/mm. Kok et al. have calculated diameters of impurity particles from published breakdown strength data, using a formula $d = (kT/2\pi\epsilon^*\epsilon_0 F_b)^{1/3}$ and correlated them with the diameters of particles existing in liquid nitrogen.[106]

A value of 100 kV/mm obtained by Kronig leads to an impurity diameter 11.0 Å, which is consistent with the lattice constant of a unit cell of P_2O_5. A diameter of 7.57 Å is obtained from Blaisse's data and may be ascribed to the unit cell length (4.5 to 7.41 Å) of hexagonal ice. There is no correlation between work functions of Cu, Au, and Pt electrodes and breakdown strengths obtained by Swan and Lewis. From these considerations, the particle

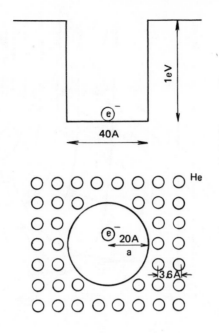

Electric Field Around A Quasi-Bubble

$$F = \frac{e}{4\pi a^2} \cong 57.5\ \text{MV/m}$$

$$(\text{cf. } \frac{1\ V}{40A} = 250\ \text{MV/m})$$

FIGURE 2.3.9. Explanation of an electron bubble.

bridge model is favored by Kok et al.[106] Medium oxidation of metal electrodes such as steel and brass would increase breakdown strength due to the formation of semiconducting oxide layers, while extreme oxidation would reduce it, due to the increase in roughness of the electrode surface.

Breakdown strengths of liquid nitrogen are shown as a function of gap length in Figure 2.3.11.[121-126] and Figure 2.3.12.,[124,125] and as a function of pressure in Figure 2.3.13.[121,124,125] According to Lehman's data, breakdown voltage is almost proportional to gap length and the square root of pressure. Fallou's data show that breakdown voltage is proportional to 0.9th power of gap length for large gap length, and that the pressure effect is more significant than obtained by Lehman. Mathes showed much lower strength in a coaxial electrode system and no pressure effect above 1.5 kg/cm² G. This may be attributable to the wide area of the electrodes and their surface irregularities.

Little is known of the breakdown process in nonuniform electric fields. According to Coelho and Sibillot,[127] electron emission into pure liquid nitrogen is abruptly selftriggered at a voltage 2.5 to 4 times higher than the threshold for emission in vacuum. Before this occurs, no current is detectable in the liquid, even when the applied voltage corresponds to a local field of the order of 10⁸ V/cm in front of the tip.

Experiments in which internal discharges taking place in a liquid-nitrogen-filled cavity indicate that the internal discharges tend to be suppressed as the pressure is increased. However, the discharges are likely to occur near the boiling temperature in spite of the pressure, as shown in Figure 2.3.14.[130]

Liquid nitrogen in a cavity is liable to overvolting, possibly because of statistical time lags. The result is a much larger discharge magnitude, as shown in Figure 2.3.15.[129,130] On the basis of the gaseous breakdown mechanism, Tanaka derived the following formulas

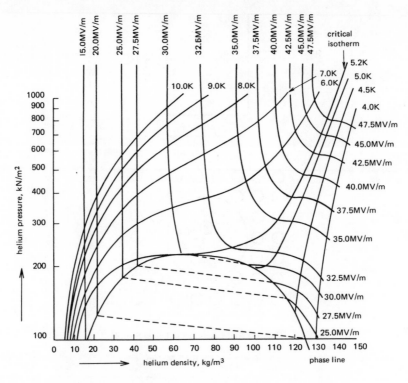

FIGURE 2.3.10.   Contour plot for helium breakdown, using brass electrodes with 1 mm gap. (From Meats, R. J., *Proc. Inst. Electr. Eng.*, 119, 762, 1972. With permission.)

$$Q_2 = (AC_b/pd)(\Delta V_g) \qquad (2.3.28.)$$

$$\Delta V_q = V_G - pdB \qquad (2.3.29.)$$

where Q_a = the apparent charge, A,B = constants, C_b = the capacitance in series with the cavity, p = the pressure, d = the cavity length, ΔV_g = the overvoltage across the cavity, and V_G = the DC breakdown voltage across the cavity.[130]

2.4. LONG-TERM CHARACTERISTICS

Cable insulation is subjected to electrical stress, heat, and various environmental hazards such as water, and in some cases, radiation. The temporal degradation of insulation due to such determining factors is an important aspect of insulation research. Mechanisms underlying aging phenomena will be discussed in the ensuing sections.

2.4.1. V-t Characteristics

The time of failure of electrical insulation after the application of voltage is called the lifetime or its incubation time. It is most important to understand what is happening within this period. Sometimes factors determining this degradation are known, although no precise mechanism is clear, but sometimes even the factors themselves are unknown. Nevertheless, a certain empirical relation between applied voltage (V) and lifetime (t) has been postulated and is widely used because of its convenience; it is as follows:

$$t = \frac{A}{V^n} = \frac{K}{F^n} \qquad (2.4.1.)$$

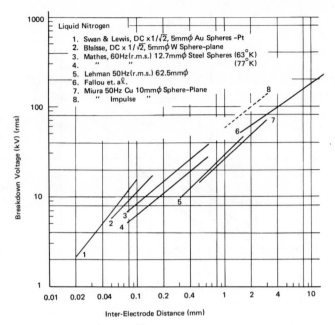

FIGURE 2.3.11. Breakdown voltage of liquid nitrogen under uniform electric field.

where A, K, and n = constants, and F = the electric stress. This relationship is linear when plotted double-logarithmically.

A critical voltage can be considered below which no degradation takes place. Consider the internal discharge; no discharge occurs at a voltage below the discharge inception voltage V_i. In this case, the following formulas apply:

$$t = \frac{B}{(V - V_i)^n} \qquad (2.4.2.)$$

$$t = C \exp\left(-\frac{V - V_i}{D}\right) \qquad (2.4.3.)$$

Since it is generally difficult to find V_i, neither of these formulas is used in analyzing V-t characteristics.

The V-t characteristics or a life curve may be theoretically treated; there are two distinct approaches. One is based on physical processes which are deemed to underlie the degradation of interest; the other is based on mathematical processes or statistical behavior of breakdown. The second treatment cannot really provide a picture of what is physically taking place. Table 2.4.1. lists various models to explain V-t characteristics.

A. Electron Avalanche Model

A formula for the growth of electrical trees due to impulse voltage application was originally developed on the basis of electron avalanche breakdown in expoxy resins.[134] Let us assume that one electron avalanche is initiated by the application of one cycle of voltage — either impulse or AC — and that this results in a deterioration of the insulation either physically or chemically. If it is assumed that an insulation specimen will fail after N cycles of the applied voltage, then we obtain

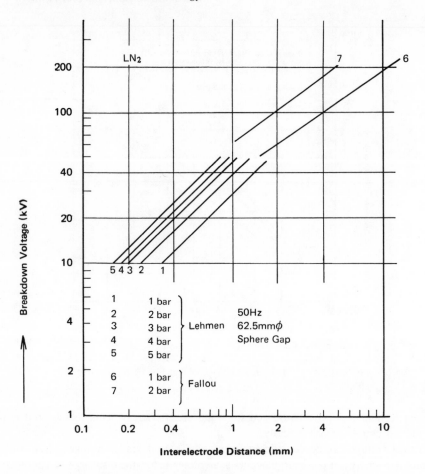

FIGURE 2.3.12.   Pressure effect on breakdown voltage of liquid nitrogen.

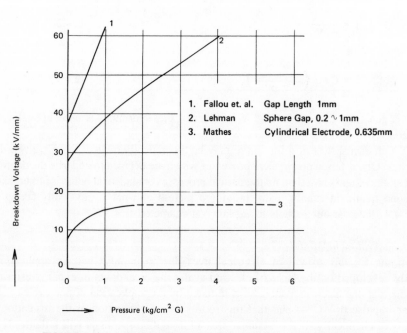

FIGURE 2.3.13.   Pressure effect on breakdown voltage of liquid nitrogen.

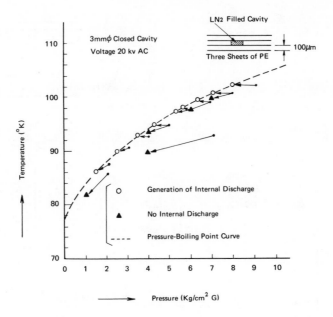

FIGURE 2.3.14. Generation of internal discharge in case of decrease in pressure.

$$K = bN = NFne \qquad (2.4.4.)$$

where K = the total energy needed for the breakdown, B = the energy supplied from one cycle of voltage, F = the applied electric field stress, n = the density of the free or conducting electrons, and e = the electronic charge. The number of free electrons multiplied by the avalanche process is given by

$$n = n_0 \exp(\alpha d) \qquad (2.4.5.)$$

where n_0 = the density of free electrons before the initiation of the avalanche process, α = the electron multiplication factor, and d = the total or some part of the inter-electrode distance. The multiplication factor α is given by

$$\alpha = D \exp(-H/F) \simeq C(F - F_i) \qquad (2.4.6.)$$

where D, H, C, and F_i are constants. Therefore the following approximate formula can be obtained:

$$n = n_0 \exp[C(F - F_i)d] \qquad (2.4.7.)$$

and consequently

$$F = \frac{A - B \log N}{d} + F_i \qquad (2.4.8.)$$

where A = $(1/C)\ell n(K/n_0 eF)$ and B = (2.3/C). By putting N = $(\omega/\pi)t$, for AC voltage, we obtain

$$(F - F_i)d = V - V_i = A' - B \log t \qquad (2.4.9.)$$

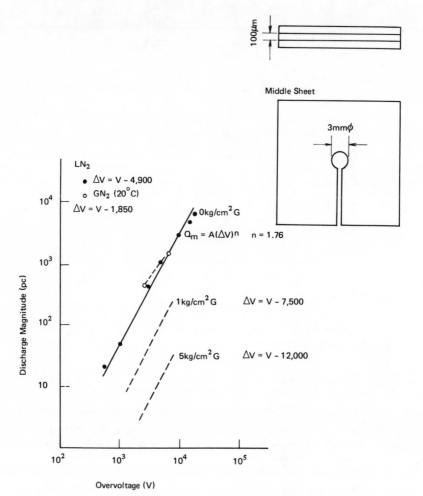

FIGURE 2.3.15. Overvoltage effects on the magnitude of internal discharge of liquid nitrogen. (From Tanaka, T., Proc. 1976 IEEE Int. Symp. EI, Montreal, IEEE, Piscataway, New Jersey, 1976, 130. With permission.)

where $A' = A - \log(\omega/\pi)$. Thus one would expect a straight line in a semilogarithmic plot.

B. Hot Electron Model

A few electrons existing in an insulator can be accelerated by the external electric field until they become hot. Such hot electrons collide inelastically with the lattice to produce irreversible changes such as the formation of radicals. Böttger derived[135] the following formula in support of this process:

$$\ln[\ln(t/t_0)] = C + \ln[(F_0/F)^\alpha - 1] \tag{2.4.10.}$$

where α is a parameter determined by the type of collision.

C. Field Emission Model

This model may apply to tree initiation.[136] It is assumed that electrons injected into the insulation from an electrode via the field emission process can contribute to tree initiation only if their energies exceed a certain critical energy. It is further assumed that they have a cumulative effect on tree initiation, i.e., the same amount of total energy can cause the same amount of damage to a dielectric.

Table 2.4.1.
A LIST OF THEORIES FOR
V-T CHARACTERISTICS

Physical theories
    Electron avalanche model
    Hot electron model
    Field emission model
    Internal discharge model

Mathematical theories
    Normal probability model
    Weibull probability model
    Modified Weibull probability model

If the critical energy value stated above is E_0, the following master equation applies for tree initiation time, t_I:

$$t_I (E - E) = C \text{ (constant)} \tag{2.4.11.}$$

Electron energy may be proportional to the field emission current and therefore be represented by

$$E = A \exp(-B\phi^{\frac{3}{2}}/F) \tag{2.4.12.}$$

where A, B = constants, F = the electric stress at the electrode, ϕ = the effective work function, which is to say the metal work function reduced by the electron affinity of the dielectric. Combining the above two equations leads to[137]

$$\ln(B\phi^{\frac{3}{2}}(F - F_0) t_I) = B\phi^{\frac{3}{2}}/F + \ln(C/A) \tag{2.4.13.}$$

where $E_0 = A \exp(-B\phi^{3/2}/f_0)$. When $F \gg F_0$, we obtain

$$\ln t_I = B\phi^{\frac{3}{2}}/F + \ln(C/A) \tag{2.4.14.}$$

D. Internal Discharge

When internal discharge is clearly involved in insulation degradation, it is most reasonable to take its accumulated effect into consideration. The total energy required for life may then be represented by

$$K = FQ = F \sum_i q_i = FNq = Fqn_0 (\omega/\pi) t \tag{2.4.15.}$$

where Q = the total charge, q_i, q = the charge of one discharge pulse, N = the total number of discharge pulses, n_0 = the number of pulses in the half cycle of AC voltage, θ = the power frequency, and t = the lifetime. The term n_0 is approximately proportional to $(F - F_i)/E_i$. The discharge magnitude q may be represented by

$$q = q_0 [(F - F_i)/F_i]^m \tag{2.4.16.}$$

This originates from the overvoltage effect of internal discharges as a consequence of statistical time lags.[129,139,140] Accordingly, we obtain[138]

$$F(F - F_i)^{m+1} \omega t = \text{constant} \tag{2.4.17.}$$

Internal discharges in a cavity in a dielectric, or adjacent to a metal electrode, can cause the eventual onset of larger rapidly growing trees. Formula 2.4.17. can apply to this case too, if the critical energy is postulated for the formation of trees.

E. Normal Probability Model[141]

If a voltage, V, is applied to N dielectric specimens each with a breakdown probability, p, then the probability P that at least one specimen breaks down is given by

$$P = 1 - (1 - p)^N \qquad (2.4.18.)$$

From statistical considerations, p at a certain time t can be obtained as

$$p = 1 - \exp\left\{ -P_0 \frac{t}{t_0} \left(\frac{V}{V_0} \right)^n \right\} \qquad (2.4.19.)$$

where p_0, t_0, V_0, and n are constants. The probability P then becomes

$$P = 1 - \left\{ \exp\left[-P_0 \frac{t}{t_0} \left(\frac{V}{V_0} \right)^n \right] \right\}^N \qquad (2.4.20.)$$

It gives a constant probability when the following equation is obeyed:

$$t = A/V^n \qquad (2.4.21.)$$

which is exactly the same as the empirical formula (Equation 2.4.1.).

F. Weibull Probability Model[142]

The Weibull distribution, based on the weakest link theory, may apply for some voltage degradation processes and may explain V-t characteristics. The two-dimensional Weibull distribution (cumulative) can be represented by

$$P(V,t) = 1 - \exp\left(-K(V - V_i)^{\frac{1}{a}} t^{\frac{1}{b}} \right) \qquad (2.4.22.)$$

where V = the applied voltage, V_i = the voltage below which no degradation takes place, t = time, and a,b = constants. This gives a constant probability when the following equation is satisfied:

$$t = A/(V - V_i)^n$$

$$n = \frac{b}{a} \qquad (2.4.23.)$$

when $V \gg V_i$, we obtain the same result with Equation (2.4.1.).

G. Modified Weibull Probability Model

Several modified Weibull distribution models have been proposed. In general, they comprise more than two Weibull distribution functions with different shape, scale, and location parameters. They may be mixed additively or multiplicatively according to the physical processes involved. V-t characteristics can be derived by computer calculation. Mixed Wei-

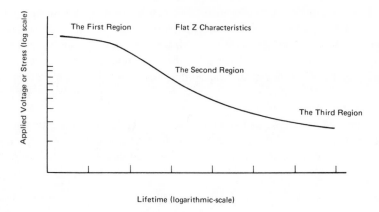

FIGURE 2.4.1.   General characteristics of life curves for insulation subjected to electric stress.

bull distributions are treated in detail in Section 2.5. Finally, it should be noted that V-t characteristics meet a flat Z curve rather than a straight line in double-logarithmic plots, as shown in Figure 2.4.1. Apparently, the curve can be divided into three regions. The experimentally obtained straight line may be a part of it, so much care should be taken when the line is extrapolated to longer times.

Both physical and mathematical approaches can be applied to degradation processes. However, they have never been theoretically correlated with each other. Such correlation is worthy of further exploration.

2.4.2. Corona Degradation[143,158]

Corona discharge deteriorates insulation and causes its final breakdown. A corona discharge is a gaseous discharge taking place within a cavity adjacent to a metal electrode, or a cavity surrounded by dielectric material. Sometimes, the terms ''partial discharge'' or ''internal discharge'' are used instead of corona discharge to describe the phenomenon.

Since an internal discharge is a gaseous discharge within a cavity (usually in a small cavity), it is readily affected by the environmental conditions. The internal surface of the cavity is subjected to physical damage or erosion, or a change in surface electrical resistivity,[143] because of the internal discharge; these effects in turn act to change the behavior of the internal discharge. In a closed cavity, the trapped gases will change with time due to the consumption of some gases such as oxygen for chemical reaction with the cavity inner surface, and the generation of various kinds of chemical species such as ozone, metastable states of molecular oxygen, atomic oxygen, and nitrogen oxides. Pressure may change too, again influencing internal discharge behavior. All these factors are responsible for the apparent characteristics of discharges in cavities, i.e., their intermittency, cessation, and resumption.

More than three mechanisms have been proposed for insulation degradation by internal discharges in insulation. They are not always inconsistent but sometimes parallel or serial. Firstly, insulation will decompose through chemical reaction with activated oxygen, this consisting of oxygen ions, excited oxygen molecules, or atoms generated by the corona discharge. It is believed that this reaction takes place with the inner surface material of the cavity, which oxidizes and produces mainly water (H_2O) and carbon dioxide (CO_2), and is at the same time eroded. This is the case with direct exposure to the discharge, because the lifetimes of the activated oxygen atoms are very short.

A second degradation process involves ozone and nitrogen oxides which have relatively long lifetimes. Ozone will diffuse into such insulation as polyethylene to react with double bonds and make ozonides, and with methylene radicals in main chains and carbonyl compounds consisting of carbonic acid. Nitrogen oxides (NO_2) will react with polyethylene to create nitric acid ester. A third mechanism of degradation has high and low energy particles, produced in the corona discharge, damaging the insulation. It is thought that the effect of high energy electrons is negative. Some effects of attack by ionic particles have been demonstrated.[146-149,155] Ultraviolet light emission from a corona discharge may change the chemical structure of insulation.

From the above discussion, the first process seems to be the most important. The following chemical reaction, associated with the formation of activated oxygen, is proposed in the case of a point-to-plane electrode system. It is assumed that the well known reaction

$$O + O_2 + O_2 \rightarrow O_3 + O_2 \tag{2.4.24.}$$

is the main process leading to the formation of O_3, that the oxygen atom involved in Reaction 2.4.24. is excited to the ${}^1D$ state, and that a small amount of water exists. Under these conditions, the following chain of decomposition can take place.

$$\left. \begin{array}{l} O + H_2O \rightarrow 2OH \\[2mm] OH + O_3 \rightarrow HO_2 + O_2 \\[2mm] HO_2 + O_3 \rightarrow OH + 2O_2 \end{array} \right\} \tag{2.4.25.}$$

This can explain the results obtained for positive corona. For negative polarity, an alternate series of chemical reactions are under consideration as follows:[157]

$$\left. \begin{array}{ll} O^- + O_2 + O_2 \rightarrow O_3 + O_2^- & \text{(A)} \\[2mm] \text{or} & \\[2mm] O^- + O_2 + O_2 \rightarrow O_3^- + O_2 & \text{(B)} \\[2mm] \text{followed by} & \\[2mm] O_3^- + O_2 \rightarrow O_2^- + O_3 & \text{(C)} \end{array} \right\} \tag{2.4.26.}$$

The increase of ozone production observed when water vapor is introduced may be due to clustering of H_2O molecules into certain negative ions — such as $NO^-{}_2$ — that are destroyers of O_3. This is based on the assumption that reactions between clusters $A^-(H_2O)_n$ and O_3 have lower rate constants than reactions between A^- and O_3. Reactions (A) and (C) are endothermic by about 1 eV and certainly occur in the high field region of the discharge near the point. In negative polarity, therefore, ozone is produced only in a small volume in the vicinity of the point, whereas in positive polarity it is produced along the entire path of the streamer from the point to the plane where atomic 0 can be produced.

One electrode system for internal discharge experiments is shown in Figure 2.4.2.; it consists of three sheets of polymer film with the middle one perforated. A specimen of the three sheets is placed between two circular electrodes. Figures 2.4.3. and 2.4.4. show two examples of failure rates vs. applied voltage in a Weibull plot. These are the results of work done in a round robin test by the Committee on Corona Degradation of Insulation Materials of the Institute of Electrical Engineers of Japan. It is apparent that the curves obtained do not precisely fit the standard Weibull distribution but indicate instead concave characteristics in Weibull plots. Figure 2.4.5. demonstrates the flat Z characteristic in life expectancy curves (median values) for the selected materials.

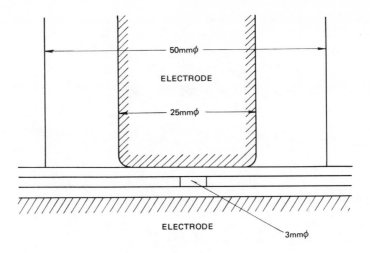

FIGURE 2.4.2. An electrode system for internal discharge experiments.

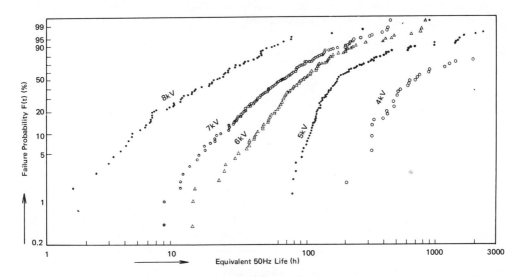

FIGURE 2.4.3. Weibull distribution of lifetimes of polyethylene subjected to internal discharges. (From Okamoto, H., Kanazashi, M., and Tanaka, T., *Deterioration of Insulating Materials by Internal Discharge,* IEEE, Piscataway, N.J., 1976, 163. With permission.)

Gas kind and pressure are also important. In a sealed cavity, oxygen is consumed in whatever mechanism is involved, as shown in Figure 2.4.6. This implies pressure change. Nitrogen seems to have no effect on pressure change. Only the cavity edge is eroded or pitted in a sealed or open cavity supplied with nitrogen. An oxygen discharge, which may be realized in an open cavity with gas flow, produces uniform erosion over the inner surface of the cavity. Figure 2.4.7. shows a marked difference between the effects of nitrogen and oxygen discharges. Pit formation in a cavity located in a thick dielectric will result in treeing and lead to its final breakdown.

Figure 2.4.8. illustrates the sequence of processes involved in internal discharges and their effects on material degradation; it is selfexplanatory.

2.4.3. Treeing

Treeing is a phenomenon whereby partial breakdown or dendritic paths progressively grow and branch into hollow channels in a solid dielectric.[165-173] They have the appearance of

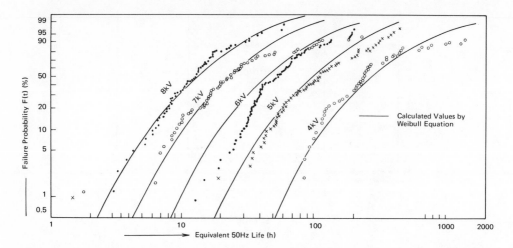

FIGURE 2.4.4.   Weibull distribution of lifetimes of polyethyleneterephthalate subjected to internal discharges. (From Okamoto, H., Kanazashi, M., and Tanaka, T., *Deterioration of Insulating Materials by Internal Discharge,* IEEE, Piscataway, New Jersey, 1976, 168. With permission.)

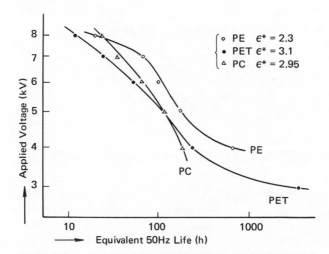

FIGURE 2.4.5.   V-t characteristics of polyethylene, poiyethyleneter-ephthalate, and polycarbonate. Power frequency: 50 Hz, 400 Hz, 1000 Hz. Failure voltage: median of 20 to 50 experimental datum. (From Okamoto, H., Kanazashi, M., and Tanaka, T., *Deterioration of In-sulating Materials by Internal Discharge,* IEEE, Piscataway, New Jersey, 1976, 169. With permission.)

trees, therefore such breakdown paths are called electrical trees, or simply trees. They can occur at rather low applied voltage but possibly at very high electric field. Tree-like discharge patterns, sometimes leading to total breakdown of the insulation used, have been observed for many years in oil-impregnated pressboard, and oil-impregnated laminated-paper cables. They propagate preferentially in a direction parallel to the internal paper interfaces and only occasionally break through layers.[174] Treeing is presently more serious in solid dielectrics such as polyolefinic polymers, rubbers, and epoxy resins, which appear to be susceptible to treeing. Once a tree starts, it is likely to cause breakdown sooner or later. They have been observed in cables, but since rubbers and resins are often pigmented or mineral-filled, the existence of tree-like channel phenomena may go unnoticed in these materials.

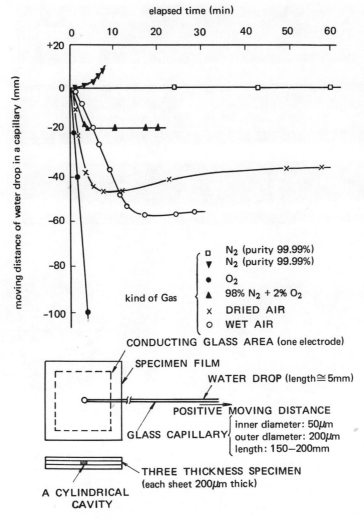

FIGURE 2.4.6. Pressure change in a cylindrical cavity due to discharges and the sample configuration. (From Tanaka, T. and Ikeda, Y., *IEEE Trans Power Appar. Syst.*, 90(6), 2702, 1971. With permission.)

A. Tree Initiation

There are two distinct time periods in treeing: the first is an incubation period during which nothing appears to happen and the second is a propagation period during which tree-like figures grow in the insulation. A tree appears immediately after the incubation period. Up to the visual appearance of a tree figure and the start of significant measurable partial discharges, no obvious effects may occur. The initiation of electrical trees from sharp point of high stress in resins, as contrasted to the propagation, is surely the least understood aspect of electrical tree development.

It may be inferred from experimental results obtained thus far, that at low stress levels where trees are not immediately observed, cumulative processes are proceeding which eventually foster conditions which initiate a tree channel. Several explanations may be offered for the initiation of an electrical tree, as indicated in Figure 2.4.9.

a. Process 1 — Tree Initiation by Mechanical Fatigue

High Maxwell compressive forces in the dielectric, caused by high electric fields at local excrescences when AC voltage is applied, produce a mechanical fatigue cracking in the polymer.

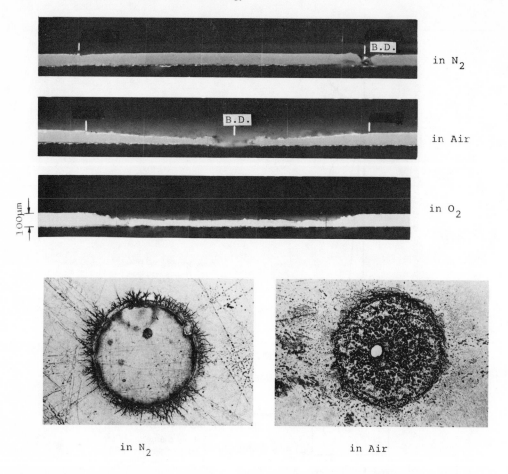

FIGURE 2.4.7.   Pit formation by nitrogen discharges and uniform decrease in thickness by oxygen discharges (polyethylene).

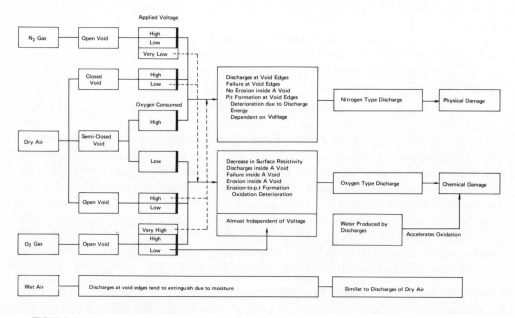

FIGURE 2.4.8.   Some explanations of effects of environmental conditions on void discharge behaviors.

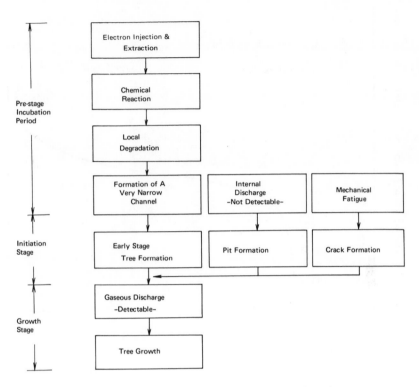

FIGURE 2.4.9.   Probable processes for electrical tree formation.

b. Process 2 — Tree Initiation by Partial Discharges

This explanation is based on the belief that small cavities can exist at tips of foreign particles, asperities, or needles, due to differential thermal expansion of the resin and the metal, or as a consequence of adsorbed or accumulated gas on the surface. It is supported by evidence.[221] Another probable cause is bombardment of the surface of the insulation by high-speed electrons, the source of the electrons being high-intensity ionization in a void, or field emission from the surface of a metal point with very small radius of curvature.

c. Process 3 — Tree Initiation by Charge Injection and Extraction

Under AC voltage conditions, some electrons will be emitted or injected for a short distance into the dielectric during the negative half cycle, the distance being limited by the declining stress away from the emitting points. If they are not trapped they will be drawn back into the point on the positive half cycle and reinjected the following cycle. In each cycle some of the electrons will gain sufficient energy to cause some resin decomposition to lower molecular weight products and gas.

As an alternative or addition to electrons injected from the electrode, it has been suggested that free electrons within the dielectric may acquire enough energy to cause inelastic collisions with lattices, leading to a molecular degradation and, finally, the initiation of a tree.[169] Electrons may be supplied from deep donors (or trap levels) via the Poole-Frenkel effect and then be accelerated in the conduction level by the external field to become hot electrons.

B. Tree Propagation

Once a tree is initiated, the growth proceeds by a series of sporadic bursts of activity. Branching becomes more frequent as the tree progresses. As the tree becomes more bushy, the rate of growth slows down until the outermost twigs approach the opposite electrode.

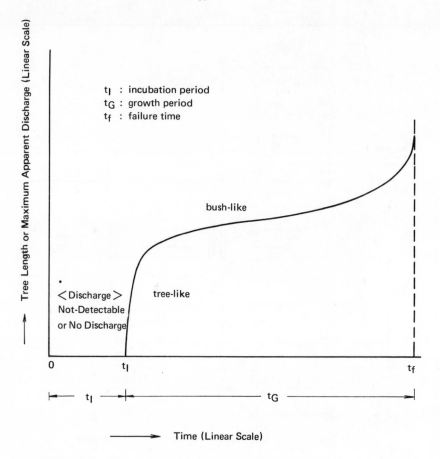

FIGURE 2.4.10.   Temporal change in tree length and its associated discharge magnitude.

At that time, if failure occurs, it occurs rapidly. This is presented schematically in Figure 2.4.10. Photographs of the two types of trees are shown in Figure 2.4.11. The discharges which take place in the narrow channels of the tree are unusual in their erratic and sporadic behavior. Perhaps one explanation is the very small volume of the discharge, coupled with its very large surface area. The buildup of static charges along the inside of the tubes, which could trap ions, has been suggested as another reason. Still another possible cause, in the case of some materials, is repolymerization of some of the gaseous decomposition products, which would narrow still further the already tiny channels.

There are two more important features which influence electric tree propagation: pressure and electrostatic shielding. The development of internal gas pressure due to the decomposition of the insulation may cause the discharges to be interrupted intermittently if the tree is tightly sealed. Pressures can be reached which will extinguish the discharge and no further discharges will occur until these pressures are reduced by diffusion through the polymer or by leakage along the needle-polymer interface. It is suggested for cross-linked polyethylene[175] that decomposition products of the cross-linking agent and voltage stabilizers, if they are present in a concentration higher than their solubility, tend to migrate into narrow tree channels. If this is the case, it could also lead to a rise in pressure.

The interaction of the fields of the branches should also be taken into consideration. After a tree develops with multiple branches, the stress may be somewhat reduced by the shielding effects of adjacent branches on each other. The effect is similar to the behavior of corona discharge from a point into a gas, which reduces the stress by mutual interaction. A bushlike tree can be compared to the multiple discharges which form an apparent glow around a point in a gas.[168]

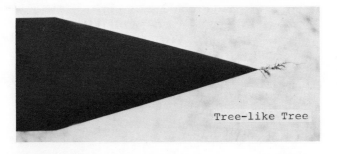

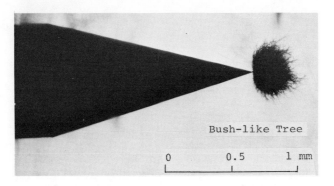

FIGURE 2.4.11. Photographs of two types of electrical trees in poly-ethylene. (Tree starts from the tip of a stell needle.)

Tree channels are generally considered to be hollow and nonconducting, having a breakdown strength of about 400 kV/mm. Conducting carbon may be found in a tree channel in epoxy resins.

In spite of much clarification of tree growth phenomena, the precise mechanism still requires further investigation. Growth of local breakdown at the tip of a tree is interpreted mainly in terms of:

1. Intrinsic breakdown at a tree tip under the intensified electric field which is probably in the range of the intrinsic breakdown strengths of polymers, i.e., $10^6 \sim 10^7$ V/cm
2. Inner surface bombardment by high energy electrons and ions, which constitutes a kind of corona degradation, and results in decomposition through thermal and chemical reaction

Injection of electrons from the gas phase into the solid phase is another possible cause. This phenomenon is observed when a dielectric is irradiated by high-energy electrons and then short circuited with a point contact. A mechanical failure due to increased internal pressure is suggested as yet another mechanism.[176]

C. Voltage and Frequency Acceleration

It is important to be able to estimate the lifetime of a given dielectric from short-time experiments or accelerated tests. Voltage aging, unlike thermal aging, may be accelerated by higher voltage and frequency. Clarification of such acceleration laws, if any exist, will shed light on the mechanism of treeing too. So far no definite acceleration law has been established, but for convenience life expectancy may be represented by the following formula:[177,178]

$$L = C/f^x F^n \qquad\qquad (2.4.27.)$$

Table 2.4.2.
INDEXES IN A SIMPLIFIED LAW FOR VOLTAGE AND FREQUENCY ACCELERATION IN VARIOUS VOLTAGE AGING

Kind of aging	n	x
Surface discharge	~ 2	Close to 1
Void discharge	$2 \sim 4$, 9	$< 1^a$
Electrical trees[b]		
Inception	$5 \sim 34$	Close to 1
Progation	—[c]	0^c
	4.3^d	0.23^d
Water trees[b]	~ 1	Close to 1

[a]   Dependent on the degree of airtightness of a cavity.
[b]   For polyethylene.
[c]   In case of gas leak.
[d]   In case of no gas leak.

where f = the voltage frequency, F = the electric stress, n and C = constants, and x = a factor less than unit. Values for n and x are tabulated in Table 2.4.2.[178] It is rather surprising that in some cases there is no acceleration due to frequency.

D. Principles of Inhibiting Tree Formation

Although there is no definite theory in respect to the precise mechanism of electrical treeing, it is certainly accepted that electrical trees start from high electric stress regions in dielectrics. A common practice is to introduce other materials called "voltage stabilizers" or additives to reduce the possibility of tree generation. Gases, oils or liquids, waxes, antioxidants, catalyst stabilizers, and mineral fillers of low hygroscopicity are all candidates for compounding agents for this purpose.

Theoretically speaking, there are four principles one may follow to inhibit treeing; although from experimental points of view, the evidence for making a distinction between the four is fragmentary. The four principles are[179-185]

* Principle I: to fill voids with some suitable materials
* Principle II: to coat the internal surface of voids with a semiconducting compound
* Principle III: to relax locally intensified electric stresses
* Principle IV: to trap or decelerate high energy electrons

Table 2.4.3. shows some examples of voltage stabilizers and their probable functions. The first and second principles apply where a tree is initiated via void discharges. The third and fourth principles are suitable where trees start without the benefit of void discharges.

It is reasonable to replace the original gases in a cavity (perhaps CO_2 or air) with some other gas or liquid, or even solid. In order to increase corona inception voltage of the entire insulation when voids are present, the substituting material should have either a higher dielectric strength or a higher permittivity than the original gases. When unsaturated hydrocarbons are used as impregnants, they may perhaps act as stabilizing, hydrogen-absorbing liquids.

Lowering the surface resistivity of a void by additives can suppress internal discharges because the void is virtually shorted out. A simple calculation of the electric field in a

Table 2.4.3.
ADDITIVES TO INHIBIT TREEING

	Applicable principles	Examples of additives or fillers
Gases	Principle I	SF$_6$ N$_2$
Liquids	Principles I and II	Silicone oil, cable oil, polybutene, paraffin oil, dimethylglycol, trimethylglycol, acetophenone, cumyl alcohol, di-tertiary-butyl-*p*-cresol, thio-bis (6-t-butyl-m-cresol)
Organic semiconducting compounds (antioxidants)	Principles II, III, IV	Diphenyl-*p*-phenylene diamine (DPPD), phenyl-d-naphthyl amine, pentachlorophenol, 4-bromodiphenyl, hexabromobiphenyl 1, 2-dihydroquinoline, 8-hydroquinoline, polymerized 2,2,4-trimethyl-1,2-dihydroquinoline, ferrocene, siloxane oligomers
	Principles II and III	M (Fe, Pb, Bi, Se, Cu, Ni)-dimethyl (or dibutyl)-thiocarbamate, N-N diethylthiocarbamoyl, 2-benzothiazoyle sulfide, bidenzothiazol disulfide
Inorganic mineral fillers	Principle III	Talc, calcinated clays, mica, silica, magnesia, calcium, carbonate

spherical cavity will show that much reduction in the internal electric field can be anticipated[183] if the inner-surface resistivity is reduced to around $10^{10}\Omega$. Various additives are available to provide insulation with such performance. When mixed with insulating materials, they bleed out to surface of insulation, including the void surfaces. There are two groups of the additives, one has low resistivity by nature, while the other exhibits low resistivity only after being subjected to corona discharges.

After mixing in an additive, a specimen needs a certain time before it will exhibit the benefits of the additives. It is known in treeing experiments as the "rest effect". It can be interpreted in terms of the time required for additives to bleed or bloom to any insulation surface. The voltage required for tree generation increases once a specimen with a needle is subjected to even a low voltage. This is probably because the concentration of additives tends to increase as applied voltage increases (possibly due to the dielectrophoretic force), or by virtue of the fact that bleeding additives become semiconducting through corona discharges, thereby causing a decrease in the electric field around the needle and the voids.

Field grading with additives or voltage stabilizers is applicable to the third principle. Voltage stabilizers are expected to bleed or bloom to asperities or contaminants with high conductivity or high permittivity, and so screen them electrostatically by forming a Faraday cage. They act to soften sharp electrode profiles and reduce electric field around them. The addition of these semiconducting materials increases the dissipation factor, which in some cases can cause thermal runaway of the insulation.

An idea is proposed to take advantage of the field dependence of the conductivity of weak electrolytes, the so-called second Wien effect. This phenomenon involves the increased dissociation of weak electrolytes into ions when the electric fields is increased, thereby producing a higher conductivity in the high-field region and therefore a reduction in the field. Conversely, in the same circumstance, these materials show a low conductivity in the low-field region, hopefully the result is a low tanδ for the overall insulation.

Principle IV is more or less related to the intrinsic breakdown process of polymers, which is based on the hypothesis that high-energy electrons can be decelerated and thermalized by the resonance trapping inherent in some chemical structures such as aromatic hydrocarbons or benzene rings. This is described in Figure 2.4.12.

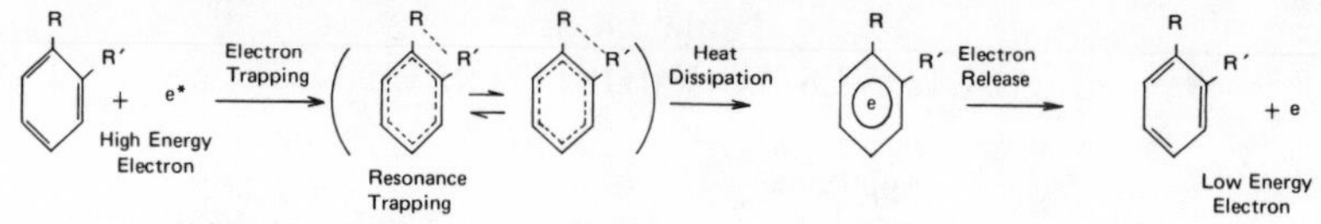

FIGURE 2.4.12.   Hypothetical process for slow down of high energy electron by resonance trapping.

Introduction of a mineral-type filler can improve material performance. It is usually added in a larger amount than is any additive. Its function can be classified under Principle III, but it seems to operate somewhat differently from additives. Its performance is interpreted in terms of space charge effects as well as inherent physical and chemical properties of inorganic fillers.[185]

Finally, it is to be noted that additives must be soluble and stable in the polymer of interest (PE, XLPE, or EPR), which means that they should probably possess fairly large hydrocarbon-like groups. In addition, they must not bring about a significant increase in tanδ.

2.4.4. Water Treeing

Phenomena and characteristics associated with water trees are described in the second volume. Here we shall discuss the possible mechanisms of water treeing, or electrochemical trees as they are sometimes called. There are two schools of thought, but both theories are still speculative.

The first is concerned with dielectrophoresis followed by some other processes such as electrostriction and thermal expansion. The second theory is developed on the basis of the chemical potentials, and seems to depend on electrostrictive (Maxwell) forces.

A. Dielectrophoresis as a Water-Collecting Process[186]

The situation shown in Figure 2.4.13. is now under consideration as a means to calculate the time-dependent distribution of impurity concentrations. This is a model for an entire cable cross section or a part thereof. It contains a particle, which may represent water or impurities in water in the cable insulation. The force upon a neutral particle is given by[189]

$$F = -4\pi a^3 \epsilon_0 \epsilon_1^* \frac{\epsilon_2^* - \epsilon_1^*}{\epsilon_2^* + 2\epsilon_1^*} \cdot \frac{V^2}{r^3[\ln(r_1/r_2)]^2} r^0 \qquad (2.4.28.)$$

where a = the radius of particle, ϵ_0 = the permittivity of free space, ϵ_1^* = the specific dielectric constant of a medium such as polyethylene or cross-linked polyethylene, ϵ_2^* = the specific dielectric constant of impurities such as water, V = the applied voltage, r_1 = the radius of conductor, r_2 = the radius of cable, r = the distance from the center of the conductor, and r^0 = the unit radius vector. The force is seen to be proportional to the square of applied voltage and therefore to be independent of the voltage polarity. The direction of the force is opposite to the radius vector and the particle moves toward the conductor or the higher electric field region.

The particle is accelerated by this force and decelerated by any collisional force. Newton's generalized equation is given as follows:

$$\frac{d^2r}{dt^2} + \frac{1}{2}\frac{dr}{dt} + \omega_0^2 r = -\frac{\alpha}{mr^3} V^2 \qquad (2.4.29.)$$

where τ = mD/dT (Einstein's relation, m = the electronic mass, and

$$\alpha = \frac{4\pi a^3 \epsilon_0 \epsilon_1}{[\ln(r_1/r_2)]^2} \cdot \frac{\epsilon_2^* - \epsilon_1^*}{\epsilon_2^* + 2\epsilon_1^*}$$

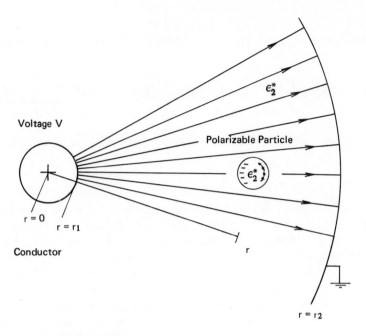

V : Applied Voltage

r_1 : Radius of Conductor or A Single Strand Conductor

r_2 : Radius of Cable

a : Radius of Particle

ϵ_1^* : Specific Dielectric Constant of Insulation

ϵ_2^* : Specific Dielectric Constant of Particle

FIGURE 2.4.13.   A cross section of a polyethylene insulated cable containing a particle as an impurity.

Since there is no inherently stable position for the particle in the case of interest, the third term of Equation 2.4.29. can be neglected. Generally in solids, the collision term is more dominant than the acceleration term, we therefore obtain:

$$v(r) = \frac{dr}{dt} = \frac{\alpha\tau}{mr^3} \, V^2 \qquad (2.4.30.)$$

The time-dependent distribution of the concentration of impurities can be determined by the balance between dielectrophoretic and diffusional particle flux rates. This is given by:

$$\frac{\partial n}{\partial t} = DV^2n - \text{div}\,(nv) \qquad (2.4.31.)$$

where D is the diffusion constant. The solution obtained in the special case in which the concentration of particles at $r = r_2$ and $t = 0$ is assumed to be a delta function.

$$n(r, t) = \overline{n}\,(r_2/r)^{1/2}\,\exp\left\{\frac{\beta}{4r^2}\left(1 - \frac{r^2(r - r_2)^2}{\beta Dt}\right)\right\}$$

$$\text{where } \overline{n} = \frac{\displaystyle\int_{r_1}^{r_2} n(r, t)dr}{\displaystyle\int_{r_1}^{r_2} (r_2/r)^{1/2}\,\exp\left\{\frac{\beta}{4r^2}\left(1 - \frac{r^2(r - r_2)^2}{\beta Dt}\right)\right\}\,dr} \qquad (2.4.32.)$$

Table 2.4.4.
TIME FOR WATER AGGREGATION BY DIELECTROPHORESIS

	r_1 (mm)	r_2 (mm)	T (K)	V (volts)	β	tc	Remarks	
Case 1	0.4	4	300	3,800	1.95×10^{-5}	4.4 years	$D = 10^{-14}$ m^2/sec	6.6 kV cable
Case 2	0.4	4	368	15,200	2.53×10^{-4}	12.4 days	$D = 10^{-13}$ m^2/sec	6.6 kV cable, accelerated
Case 3	0.2	1	300	6,000	8.03×10^{-5}	1.5 days	$D = 10^{-14}$ m^2/sec	dielectrophoresis

The characteristics of the obtained time-dependent distribution of particle concentration can be described as follows. The distribution is essentially determined by the exponential factor in Formula 2.4.32. When the reduced time ($\beta Dt/r^4_2$) is smaller than 0.106, there is 1 minimum point at $4 \le (3/4)r_2 = r_m$, and 1 maximum point at $r > r_m$. They converge on r_m. When the reduced time becomes larger than 0.106, the distribution has no minimum and maximum points and therefore the particles become densely populated near $r = r_1$.

Various parameter values must be determined for the calculation of Formula 2.4.32. Water is chosen as the impurity. The specific dielectric constant ϵ^*_2 of water is 80.36, which is larger than that of any other ordinary impurity; thus water is most influenced by the dielectrophoretic force as Formula 2.4.28. shows. Other impurities, if any, must move with the water. The size of the water is not fixed; it will be determined by the amorphous or porous (or microvoid) parts of the polyethylene. A reasonable value for the radius of a particle might be 500 Å. This is similar to the radius of the smallest water trees. Diffusion constant of water in polyethylene may be in the range of 10^{-14} m^2/sec at room temperature. Examples of calculated results are shown in Table 2.4.4. and Figure 2.4.14. The time, t_c, is a measure for the time to steady state, and therefore is considered to be the induction period for water trees. This process is substantiated by a model experiment.[186]

B. Processes Subsequent to Dielectrophoresis[187]

The following mechanisms are based on experimental results of residual strain in water tree regions as shown in Figure 2.4.15. and some evidence for the formation of CH_3 radicals which is related to the scission of polymers.

Unlike electrical trees, water trees consist of discrete chains of microvoids (order of 100 μm) which align roughly in a direction parallel to electric lines of force as shown in Figure 2.4.16. Two possible processes can be considered to explain the formation of these chains:

1.   The force of thermal expansion of water in voids due to a selective heating process
2.   An electrostrictive force

Consider a slim microvoid (not a perfect spherical) in polyethylene, which is filled with water, as shown in Figure 2.4.16.(A). If only the water is heated by a selective heating process (dielectric or joule heating), the microvoid will change shape by expansion of the water, becoming more spherical. Compressive and tensile stresses are forced on the polyethylene in the short-radius and long-radius directions, respectively.

Submicrovoids can be formed in the region subjected to tensile stress, i.e., region A. A number of submicrovoids are created by repeated expansion and contraction of the water, which aggregate into a single microvoid, so that a new microvoid is formed in line with the original one. The repetition of this process results in the formation of chains of microvoids as shown in Figure 2.4.16.(B), which is the actual shape of water trees. A chain of microvoids is probably converted into a filament-like path in its final stage, corresponding to the trunk of a water tree.

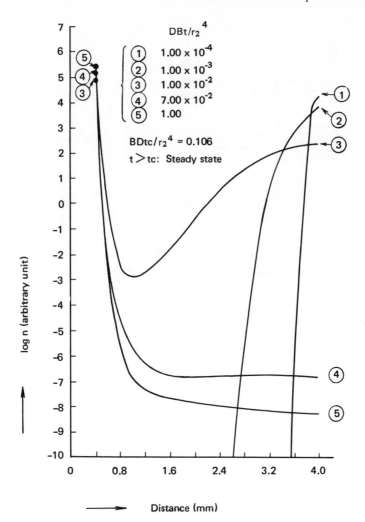

FIGURE 2.4.14. An example of theoretical time-dependent distribution of particle concentration. (Case 1 in Table 2.4.4.) (From Tanaka, T., Fukuda, T., Suzuki, S., Nitta, Y., Goto, H., and Kubota, K., *IEEE Trans. Power Appar. Syst.*, 93(2), 695, 1974. With permission.)

The direction of branch growth of water trees is thought to be determined preferentially by such factors as (1) the electric lines of force, (2) the long radius direction of a slim void, and (3) the probability for the existence of a mechanically weak point, i.e., an amorphous region among crystalline regions. Inherently existing voids may be responsible for starting water trees.

The rate of temperature rise of the entrapped water by selective heating is given by:

$$dT/dt = \omega \epsilon_0 \, \epsilon_2^* E^2 \, \tan\delta / \alpha_2 C_{V_2} \qquad (2.4.33.)$$

provided there is no heat leak, where ω = the power frequency, ϵ_0 = the permittivity of free space, ϵ_2^* = the specific dielectric constant of water, E = the electric field strength, α_2 = the density of water, and C_{V_2} = the specific heat at constant volume. The value of the rate is estimated to be 0.905°C/sec, when $\epsilon_0 \epsilon_2^*$ = 7.12 × 10^{-10} F/m, $\tan\delta$ = 0.71, C_{V_2} = 4.20 × 10^{-6} J/m² degree, ω = 314 sec⁻¹ and E = 10^7 V/m. Heat is actually dissipated through polyethylene insulation and therefore the thermal time constant is given by the formula

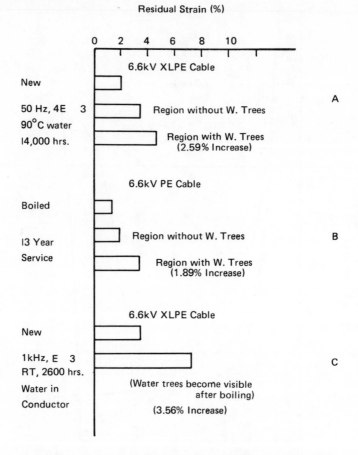

FIGURE 2.4.15. Residual Strain in XLPE and PE Cable Insulation. (From Tanaka, T., Fukuda, T., and Suzuki, S., *IEEE Trans. Power Appar. Syst.*, 95(6), 1895, 1976. With permission.)

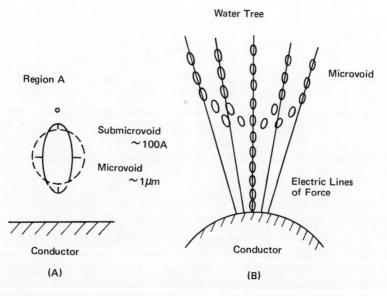

FIGURE 2.4.16. A model of water tree growth. (From Tanaka, T., Fukuda, T., and Suzuki, S., *IEEE Trans. Power Appar. Syst.* 95(6), 1895, 1976. With permission.)

$$\tau_{th} = \alpha_1 C_{V_1} d^2 / K_1 \qquad (2.4.34.)$$

where α_1, C_{V_1}, d, and K_1 are the density, specific heat, insulation thickness, and thermal conductivity of polyethylene, respectively. The value of the thermal time constant is obtained as 8.03×10^2 sec, (approximately 13 min), when $\alpha_1 = 9 \times 10^2$ kg/m³, $C_{V_1} = 2.23 \times 10^3$ J/kg degree, $K_1 = 2.5 \times 10^{-1}$ J/m·sec·degree and d = 10^{-2}m. Since the volume expansion coefficient is about 2×10^{-4} at room temperature, a 10°C temperature rise leads to 0.13% stretching of the large radius.

This means a comparatively large elongation near the tip of the microvoid, indicating the probable formation of submicrovoids near the microvoid. The volume expansion coefficient of water is 2×10^{-4} degree^{-1} and the compression coefficient of water is 4.23×10^{-5} (kg/ cm²)$^{-1}$ at 20°C in the pressure range of 0 to 500 kg/cm². When the temperature of water is raised by ΔT adiabatically, the pressure rise ΔP turns out to be $4.73 \Delta T$, assuming the volume remains constant. A 10°C rise in temperature results in pressure rise of 47.3 kg/ cm².

An alternative model is based on the electrostrictive force on a void filled with water. The force on a water-filled void, as shown in Figure 2.4.17.(A), can be calculated from the formula

$$P = 1/2E \cdot (\vec{D} \cdot n_1) + 1/2\vec{E} \times (\vec{D} \times \vec{n}) + \frac{\epsilon_0}{2} m \frac{d\epsilon^*}{dm} E^2 \cdot \vec{n} \qquad (2.4.35.)$$

where P = the pressure on an interface, $\vec{E}$ = the electric field, $\vec{D}$ = the dielectric flux density, $\vec{n}$ = a unit vector perpendicular to the interface, m = the density of materials, ϵ_0 = the permittivity of free space, ϵ^* = the specific dielectric constant.

We have to add a surface tension term to this equation. Using Onsager's formula and the Clausius-Mossotti formula for the dependence of dielectric constants of water and polyethylene, respectively, we obtain

$$P_1 = \frac{\epsilon_0}{6} \left(\frac{3\epsilon_1^*}{\epsilon_2^* + 2\epsilon_1^*} \right)^2 \left\{ 3 \left(\epsilon_2^* - \epsilon_1^* \right) \frac{\epsilon_2^*}{\epsilon_1^*} + \left(\epsilon_2^* - 1 \right) \left(\epsilon_1^* + 2 \right) \left(\frac{\epsilon_2^*}{\epsilon_1^*} \right)^2 - 3 \left(\epsilon_2^* - 1.8 \right) \right\} E_0^2 - 2\alpha/r$$

$$\simeq \frac{3}{2} \epsilon_0 \left\{ 3\epsilon_1^* + (\epsilon_2^* - 1) (\epsilon_2^* + 2) \right\} E_0^2 - 2\alpha/r = 1.44 \times 10^{-10} E_0^2 - 0.06/r \qquad (2.4.36.)$$

where P_1 = the pressure parallel to the direction of the electric field, E_0 = the average electric stress in V/m, and r = the radius of the water in meters. The estimated stress due to the electrostriction, including surface tension, is depicted by the solid line in Figure 2.4.18. where it is plotted against the electric stress with the radius of the water-filled void as a parameter. It is realized that surface tension is significant when the void diameter is small. The dashed lines in Figure 2.4.18. show the mechanical stress due to the expansion of void water vs. electric field characteristics. Mechanical stress corresponding to the elastic limit for polyethylene is around 20 kg/cm². The mechanical stress corresponding to an experimental value — 3% increase in residual strain — is 97 kg/cm² and that of 0.9% is 73 kg/cm², both estimated from a hysterisis curve of the stress-strain characteristics of polyethylene. From the above estimation, it is concluded that the mechanical stress generated with the formation of water trees is of the order of several tens of kg/cm².

As far as the absolute value of mechanical stress is concerned, it would be reasonable to consider the polyethylene subjected from the existing voids to mechanical stresses beyond its elastic limit. The direction of this stress is parallel to the electrical lines of force and outward from a single void. The void can expand as a consequence of this force and be constrained to a somewhat ellipsoidal shape. It would seem to be difficult to explain by this process how the new void is created near the tip of the original void.

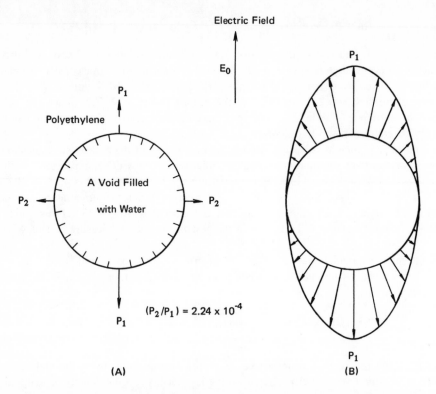

FIGURE 2.4.17. Electro-strictive force on a water-filled void embedded in polyethylene: (A) water-filled void (B) stress distribution (schematic). (From Tanaka, T., Fukuda, T., and Suzuki, S., *IEEE Trans. Power Appar. Syst.*, 95(6), 1896, 1976. With permission.)

If the temperature rise is more than 3°C at about 40 kV/mm, the stress generated by water expansion would affect water tree formation. This process can explain the creation of void chains, as described before. This force acts in such a way that ellipsoid-like voids tend to become spherical. One can imagine that both kinds of force are involved in water tree formation.

C. Water Pressure Due to Change in Chemical Potential[189]

The chemical potential which explains the diffusion of water and the growth of water-filled voids in insulation can be expressed in the presence of electric field by

$$\mu = \mu_0 - 1/2\,\epsilon_0 E^2 \left(\frac{\partial \epsilon}{\partial \rho}\right)_T \qquad (2.4.37.)$$

where μ_0 = the chemical potential in thermal equilibrium and ρ = the density of water in the insulation. Water moves toward the region of lower chemical potential, and therefore a field assisted diffusion is to be expected. This seems to be related to dielectrophoresis as described before, but so far the relation remains unclear.

The pressure of water in a microvoid (due to mechanical stress in the insulating material) can be derived from Equation 2.4.35. as follows

$$P_e = \frac{3}{2}\,\epsilon_0 \left(1 + \frac{1}{3}\,\rho_2 K_2 \left|\frac{3\epsilon_1^*}{\widetilde{\epsilon_2^* + 2\epsilon_1^*}}\right|^2\right) E_0^2 - \frac{2\alpha}{r} \qquad (2.4.38.)$$

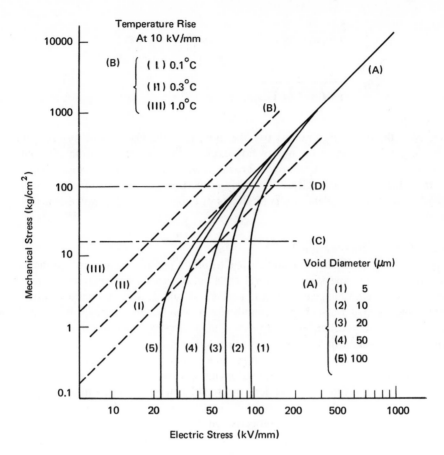

Temperature Rise At 10 kV/mm

(B) $\begin{cases} (\,\text{I}\,)\;0.1^\circ C \\ (\,\text{II})\;0.3^\circ C \\ (\text{III})\;1.0^\circ C \end{cases}$

Void Diameter (μm)

(A) $\begin{cases} (1)\;\;\;\;5 \\ (2)\;\;\;10 \\ (3)\;\;\;20 \\ (4)\;\;\;50 \\ (5)\;100 \end{cases}$

(A) Estimated stress due to electro-striction and surface tension.

(B) Estimated stress due to the expansion of void water.

(C) Elastic limit for polyethylene.

(D) Stress for 3% increase in residual strain estimated from the hysteresis curve of stress-strain characteristics of polyethylene.

FIGURE 2.4.18.   Relation of electrical and mechanical stresses in a water-filled void. (From Tanaka, T., Fukuda, T., and Suzuki, S., *IEEE Trans. Power Syst.*, 95(6), 1896, 1976. With permission.)

where ρ_1 = the density of water, $K_2 = 7.9 \times 10^{-2} m^2/kg$, ($\epsilon_1^* = 1.8 + K_2\rho_2$) and ϵ_2^* = the complex specific permittivity ($\epsilon_2^* = \epsilon_2^* - j\sigma/\omega\epsilon_0 = 81 - j1500$ at 50 Hz). According to this theory, a microvoid can grow when $P_e > 0$. This equation is comparable to Equation 2.4.36. As the microvoid grows, it tends to be ellipsoidal in shape due to the induced force in the direction of the electric field. The electric field strength is increased at the ends of the ellipse because of its geometry. From the criterion $P_e > 0$, even smaller voids are expected to grow. Thus the theory can interpret how chains of voids are formed.

It is important to calculate void size at intervals after growth starts. The following formulas are derived on the bases of Navier-Stokes equation and the diffusion equation with respect to the chemical potential for water flowing into a microvoid:

$$r = r_0 \exp(t/\tau) \qquad\qquad (2.4.39.)$$

$$\tau_0 = \frac{4}{3} \frac{\eta}{\rho_2 K_2 \epsilon_0 E_1^2} \qquad (2.4.40.)$$

$$E_1 = \frac{E_0}{1 + \left(\frac{\tilde{\epsilon}_2^*}{\epsilon_1} - 1\right)\vec{n}} \qquad (2.4.41.)$$

where r_0 = the initial void radius, t = the time, η = the viscosity of insulating material, E_1 = the field strength in the microvoid, and n = the eccentricity of an ellipsoidal void. The term τ_0 is a measure of microvoid growth time. It turns out to be closely related to the growth of water trees. It is clearly inversely proportional to the square of the electric field strength, which may explain observed effects of temperature and power frequency on tree growth.

2.4.5. Thermal Degradation[52,53,190]

Aging of organic insulating materials has its origin in chemical changes in their material structure. Generally speaking, chemical reaction increases as temperature is raised, thus aging is accelerated. Change in physical, mechanical, and electrical properties follow. Certain microscopic parameters, such as the number of broken bonds, assuredly follow a chemical reaction; this is the so-called Arrhenius's relation. Macroscopic parameters of special interest, such as tanδ, breakdown voltage, mechanical strength, and elongation should be obtained from the microscopic parameter, but in general, such correlations are difficult to obtain. It is even more difficult to determine life expectancy of cable insulation by this method because life is a complicated function of various macroparameters. Consider oil-impregnated paper for instance; mechanical strength, especially bending resistance, decreases much faster with thermal aging than does any other property, yet the paper will retain sufficient breakdown strength, in spite of the decrease in bending resistance, provided it does not induce cracking. On the other hand, breakdown may be initiated by gas generation which may accompany the decrease in mechanical strength. Accidental decrease in oil pressure would make this more probable. Thus, it is important to identify a representative macroscopic parameter for life expectancy.

Figure 2.4.19. illustrates a simple concept for the thermal aging processes. There will be a certain change in a microscopic parameter, p, at a particular temperature over a certain period of time. One may usually measure a change in a macroscopic physical quantity, q, which reflects p. In the simplest case, the change in p can be identified with the rate process for a single activation energy. Generally, more than one activation energy for chemical reaction is involved. In the aging of paper, for example, these correspond to the scission of main chains of cellulose, the decomposition of cellulose itself, the decomposition of hemi-cellulose, and so on. Now the chemical reaction of the microscopic parameter, p, is represented by:

$$\frac{dp}{dt} = f(p) \exp\left(-\frac{E}{kT}\right) \qquad (2.4.42.)$$

where p = a quantity related to material structure (the number of broken bonds of main chain and the number of produced dipoles, for example), t = the time, T = the temperature, k = Boltzmann's constant, E = the activation energy for chemical reaction, and f(p) = function of p. We then obtain

$$\int_{p_0}^{p} \frac{dp}{f(p)} = \int_{0}^{t} \exp\left(-\frac{E}{kT}\right) dt = \theta = \tau_0 \exp\left(-\frac{E}{kT_0}\right)$$

$$(2.4.43.)$$

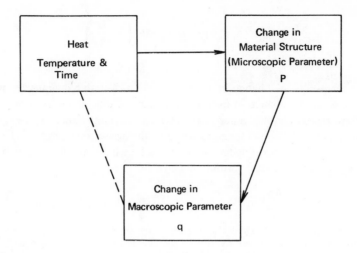

FIGURE 2.4.19. Relation between thermal aging and measurable quantities.

where p_0 = the reduced time for a given process, and τ_0 = the equivalent time at temperature T_0. A relation between the two quantities, p and q, is given by:

$$q = g(p) \qquad\qquad (2.4.44.)$$

Equation 2.4.43. can be solved, assuming the degree of reaction as follows:

- The zero order of reaction: $f(p) = A$

$$(p - p_0)/A = \theta \qquad\qquad (2.4.45.)$$

- The first order of reaction: $f(p) = Bp$

$$\{\ln(p/p_0)\}/B = \theta \qquad\qquad (2.4.46.)$$

- The second order of reaction: $f(p) = Cp^2$

$$(1/p_0^2 - 1/p^2)/C = \theta \qquad\qquad (2.4.47.)$$

Generally speaking, it is difficult to stipulate the function g(p), but in some cases the quantity q is nearly proportional to the quantity p. The dielectric tangent $\tan\delta$ of cellulose paper, for example, may correspond to produced ions and to dipoles such as water, carbon monoxide, aldehyde radicals, and carboxyl radicals.

Once any rate-determining formula is established under a certain condition, it will be possible to estimate life expectancy of a given material from an accelerated experiment at elevated temperatures. It is necessary to choose a suitable value to represent the ultimate condition ($\tan\delta = 2.8\%$ at 100°C for example) beyond which the insulation is considered no longer of value. Much care should be taken regarding the temperature range over which the accelerated aging experiment is carried out because different chemical reactions may often be involved in different temperature ranges. Experiments at too high a temperature may invalidate this extrapolation method, although they may give quick answers.

For convenience we will assume that q = p. Let q_m be the ultimate value determining life, then the isothermal aging process is governed by:

$$\int_{q_p}^{q_m} \frac{dq}{f(q)} = t_{m_i} \exp\left(-\frac{E}{kT_i}\right) = \tau \exp\left(-\frac{E}{kT_0}\right)$$

$$(2.4.48.)$$

The first term represents the change in a function of a measurable quantity, q, as it affects insulation lifetime, or the change in itself (such as $\tan\delta$ in case of the zero order reaction). The second term represents the reduced time at a certain temperature, T_i. Several plots of the reduced time against different temperatures give the activation energy, E. Extrapolation to lower temperature, i.e., operating temperature, T_0, for a particular insulation, will disclose the lifetime, τ. The mathematical formula is given by

$$\tau = \frac{\exp(-E/kT_i)}{\exp(-E/kT_0)} \, t_{mi}$$

$$(2.4.49.)$$

Cable insulation undergoes load cycling, i.e., the temperature change with load. It is convenient to use an equivalent time, τ_0 in place of τ.

$$\tau_0 = \frac{\int_0^t \exp(-E/kT)dt}{\exp(-E/kT_0)}$$

$$(2.4.50.)$$

where the temperature, T, is a function of time. If the load change with time is known, τ_0 can be calculated. The equivalent time, τ_0, should be compared with the lifetime, τ.

Examples of measured values of apparent activation energies for degradation are shown in Table 2.4.5. It should be noted that the rate of property change or activation energy for degradation depends on environmental conditions. Moisture, copper ions, and oxygen will accelerate degradation.

Thermal degradation of rubbers and plastics is generally considered to be governed by the following chains of chemical reactions.

-    Initiation reaction

$$RH + (\text{Heat, Oxygen, Light}) \rightarrow R\cdot + \cdot H$$

where R = a saturated hydrocarbon.

-    Growth reaction

$$R\cdot + O_2 \rightarrow ROO\cdot$$

$$ROO\cdot + RH \rightarrow ROOH + R\cdot$$

$$ROOH \rightarrow RO\cdot + \cdot OH$$

$$2ROOH \rightarrow RO\cdot + ROO\cdot + H_2O$$

$$\left. \begin{array}{c} RO\cdot \\ ROO\cdot \end{array} \right\} + RH \rightarrow Products$$

-    Cessation reaction

$$R\cdot + R\cdot \rightarrow R\!-\!R$$

$$R\cdot + ROO\cdot \rightarrow \text{Stable Compound}$$

$$ROO\cdot + ROO\cdot \rightarrow \text{Stable Compound}$$

Table 2.4.5.
VARIOUS ACTIVATION ENERGIES FOR THERMAL DEGRADATION

Cellulose Paper

Physical quantity	Activation energy for degradation (eV)	
	In vacuum or oil	In air
Tensile strength	0.96, 1.06	$1.35 \sim 1.44$ (1.10, 1.33) (1.00, 1.10, 1.30)
Bending resistance	1.25	—
Tearing strength	0.92	—
Rupture strength	—	1.10
Weight decrease	—	(1.13, 1.16, 1.00, 1.09)[a] 0.65
Rate of generation of decomposition gases	(1.46, 1.72), $2.13 \sim$ 2.18 1.71, (0.96, 0.78) (1.82, $1.07 \sim 1.39$)	
Average degree of polymerization	1.29, 0.96	1.29, (0.99, 0.80)
Hygroscopicity	—	0.97
tanδ	(1.14, 0.7), 0.96, 0.64	—
Electrical conductivity	(0.85, 0.39)	

Other Cellulose Products

Tensile strength	Refined cotton	0.67
	Biscose rayon	0.69
Bending strength	Ray 100%	0.83
	Sulfite paper	0.86
Weight decrease	Cotton	1.40
	Chemistry filter paper	1.35
	Bleached linen paper	1.35
	Manila cable paper	1.38
	Bleached silk cloth	1.43

Miscellaneous

Weight decrease	Mica paper polyester	$0.8 \sim 1.0$
	Mica paper epoxy	$0.9 \sim 1.4$
	Class B varnish	$0.9 \sim 1.0$

Note: ( ): Values obtained in one series of experiments. Different values in a parenthesis represent activation energies in different temperature ranges.

[a] The four values are for α-cellulose, hemicellulose, lignin, and Douglas fir sawdust, respectively.

Polymers represented by R exhibit their own specific reactivity with oxygen and their own degree of dissociation initiated by heat and light, therefore different polymers are characterized by different oxidation and thermal aging behaviors. Maximum permissible operating temperatures are more or less determined on the basis of this thermal aging process.

2.4.6. Radiation Degradation[191-193]

Power cables as well as instrumentation and control cables in the neighborhood of a power reactor are subject to high energy radiation from γ-rays and fast and slow neutrons. When organic substances are irradiated with fast neutrons, hydrogen nuclei recoil and additional γ-rays are induced via the (n,γ) reaction by slow neutrons. Induced γ-rays are not absorbed in thin organic materials and therefore their effect can be neglected. When polyethylene was exposed to a thermal neutron flux of 10^7 n/cm^2 and the radiation of the accompanying fast neutrons and γ-rays in the experimental reactor BEPO in Harwell, England, the equivalent γ-ray absorption dosage was found to be approximately 5×10^7 rad. The procedure required an entire day in the reactor.

The organic insulating materials of cables and other equipment are exposed to radiation having a γ-ray equivalent dose rate of $10 \sim 160$ R/hr in the housing vessels of a light-water reactor. The total radiation dosage will be at least 5×10^7 for the 40-year lifetime of the reactor. According to IEEE Standard 383,[194] the materials should withstand the 40-year-long dosage plus the dosage accompanying loss of coolant accident (LOCA). Much investigation has been made of the effects of radiation with high dose rate on various organic insulating materials. These are accelerated tests which make significant reduction in the time of experiments. But a degradation process may depend on dose rate even for the same total dose. A dose-rate effect is quite probable, especially where a degradation process is controlled by the diffusion of oxygen. Furthermore, in a reactor housing vessel, insulating materials are exposed to not only radiation but also heat and moisture. It is suspected that the actual degradation process may be much more complicated than so far surmised and that the three parameters are involved simultaneously.

A. Fundamental Processes

Irradiation will cause both the scission of main chains and cross linking among molecules simultaneously. Either appears macroscopically according to the probability it takes place. Polymers with $(-CH_2-CHR-)_n$ structure are of cross-inking type, while polymers with $(-CH_2-C(CH_3)R-)_n$ structure are of disintegrating type. Polyethylene, polystyrene, polyamide, natural rubber, and silicone resin belong to cross-linking polymers, while polyisobutylene, polymethylmethacrylate, polytetrafluoroethylene, poly(monochlorotrifluoro)ethylene, and cellulose are among disintegrating polymers.

Change in properties of polymers is closely related to their chemical structures. Polymers can be ranked in order of their stability against radiation, as shown in Figure 2.4.20.[195] This order is determined from changes in mechanical and electrical properties and the quantity of gases generated when the polymers were irradiated in a reactor.

Polystyrene is comparatively stable and its mechanical properties hardly change up to an absorption dose of 10^9 rad. At the same dose, polyethylene and nylon become hard. In contrast, urena and phenolic resins will decompose and become fragile. Polymethylmethacrylate, polytetrafluoroethylene, and cellulose exhibit significant reduction in mechanical properties at an absorption dose of 10^8 rad.

When subjected to irradiation, polyethylene produces free radicals such as alkyl radicals $(-CH_2-CH-CH_2-)$, allyl radicals $(-CH_2-CH-CH=CH-CH_2-)$, and radicals with conjugated double bonds $(-CH_2-CH-(CH=CH)_n-CH_2-)$; double bonds such as vinylidene bonds ($> O = CH_2$) in side chains, vinyl bonds $(-CH=CH_2)$ in ends of main chains, and transvinylene bonds $(-CH=CH-)$ in main chains; gases such as hydrogen gas and low-grade hydrocarbons are also produced.

Irradiation in air induces oxidation processes resulting in the formation of different radicals as shown below:

$$-CH_2-CH_2- + \text{irradiation} \rightarrow -\overset{\bullet}{C}H-CH_2- + H\cdot$$

$$-CH-CH_2 + O_2 \rightarrow -\underset{\underset{O_2^{\bullet}}{|}}{CH}-CH_2-$$

$$-\underset{\underset{O_2^{\bullet}}{|}}{CH}-CH_2- \rightarrow -C\overset{\displaystyle O}{\underset{\displaystyle H}{\diagup}} + \overset{\bullet}{O}CH_2-$$

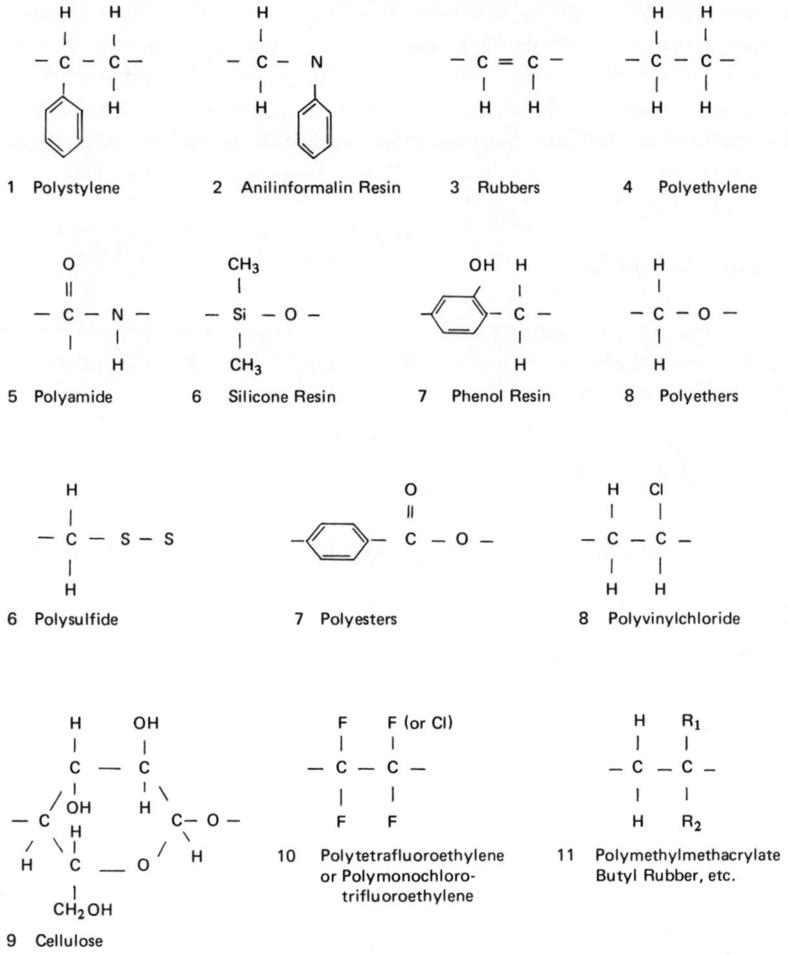

FIGURE 2.4.20. Order of radiation stability of some polymers and their chemical structures.

Here carbonyl radicals are formed. Polyvinylchloride and polypropylene are cross linked by irradiation in vacuum, while they are disintegrated by irradiation in air. Polyethylene is cross linked even by irradiation in air, but the rate of cross linking is slower in air than in vacuum. In the radiation oxidation process, oxygen diffuses from the polymer surface into the inside to react with free radicals induced uniformly by irradiation. That is to say, the process is governed by the diffusion of oxygen as well as by the generation of free radicals.

2.5. STATISTICAL ANALYSIS METHODS[196-206]

There are many probabilistic distributions to evaluate reliability such as the binominal distribution, the Poisson distribution, the normal or Gaussian distribution, the logarithmic normal distribution, the exponential distribution, the Gamma distribution, the Weibull distribution, and the extreme distribution. Of these the Weibull distribution is now widely accepted to describe the statistical behavior of cable insulation with respect to failure. The distribution is an extended form of the exponential distribution, or theoretically the third type of the extremes distribution. It was first applied to the lifetime of steel balls by Waloddi Weibull in Sweden.[197]

Two-dimensional Weibull distribution is often used to describe voltage failure of insulation with time because it is a convenient way of deducing V-t characteristics. The three dimensional Weibull distribution is sometimes used in order to include volume effects. The Weibull distribution is considered to be in "fairly good" agreement with experimental results, but in reality exhibits some deviation from them which is sometimes significant. Modifications have been made of the Weibull distribution to cover such deviations in the belief that the Weibull distribution is still applicable to a single phenomenon taking place in specimens with exactly the same electrode system.

2.5.1. Weibull Distribution
A. Derivation of the Weibull Distribution Function

Consider a system with a failure probability, f(t), at time, t. The term t is not necessarily time, and therefore the above restriction will be removed later; then f(t) is called the failure probability density function. We obtain:

$$\int_0^\infty f(t)dt = F(t) + R(t) = 1 \qquad (2.5.1.)$$

where F(t) = the cumulative failure distribution function and R(t) = the reliability. Both can be defined as follows

$$F(t) = \int_0^t f(t)dt \qquad (2.5.2.)$$

$$R(t) = \int_t^\infty f(t)dt \qquad (2.5.3.)$$

from which we obtain

$$f(t) = \frac{dF(t)}{dt} = -\frac{dR(t)}{dt} \qquad (2.5.4.)$$

We define the instantaneous failure rate as

$$\lambda(t) = \frac{f(t)}{R(t)} = -\frac{dR/dt}{R} \qquad (2.5.5.)$$

Integrating Equation 2.5.5., we obtain

$$R(t) = \exp\left\{-\int_0^t \lambda(t)dt\right\} \qquad (2.5.6.)$$

The cumulative failure distribution function is then determined by

$$F(t) = 1 - \exp\left\{-\int_0^t \lambda(t)dt\right\} \qquad (2.5.7.)$$

This is characterized by such an interesting relation as

$$\{1 - F(t)\}^m = \exp\left\{-m\int_0^t \lambda(t)dt\right\} \qquad (2.5.8.)$$

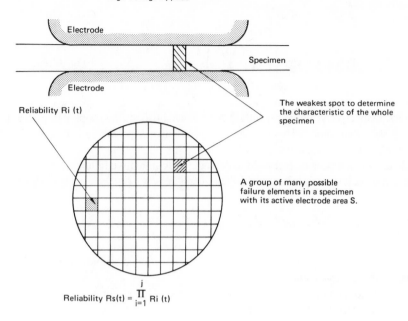

FIGURE 2.5.1. Electrical breakdown represented by the weakest link model or the serial system.

Consider now an insulating film with finite electrode area as shown in Figure 2.5.1., which is subjected to voltage. Breakdown at any spot inside the electrode area means the total failure of this film, or its life when time is involved. The film has many possible breakdown spots but is characterized only by a single spot, i.e., its weakest spot. This is certainly represented by the weakest link model, called the serial system in reliability engineering terms, because the reliability $R_s(t)$ of the whole system is obtained by taking the product of the individual reliabilities $R_i(t)$ of all possible failure elements of the system as follows:

$$R_s(t) = \prod_{i=1}^{j} R_i(t) \tag{2.5.9.}$$

This phenomenon is often explained by the scission of a chain comprising many rings, where the scission of the weakest ring determines the life of the chain.

Accordingly, the probability of nonfailure of the film specimen, $[1 - F_m(t)]$, is equal to the probability of the simultaneous nonbreakdown of every spot. Thus we have

$$1 - F_m(t) = \{1 - F(t)\}^m \tag{2.5.10.}$$

From Equations 2.5.8. and 2.5.10. we obtain

$$F_m(t) = 1 - \exp\left\{-m\int_0^t \lambda(t)dt\right\} \tag{2.5.11.}$$

We must now specify the function $\int_0^t \lambda(t)dt$. The only necessary general condition is that the function be a positive, nondecreasing function, vanishing at $t = 0$.[197] If t does not represent time, the lowest value need not necessarily be zero but perhaps some value t_L. The simplest function to satisfy this condition is $(t/t_0)^{1/b}$ where $b < 1$. Thus we put

$$\int_0^t \lambda(t)dt = (t/t_0)^{1/b} \tag{2.5.12.}$$

$$F(t) = 1 - \exp\left\{ - \left(\frac{t}{t_0}\right)^{1/b} \right\} \tag{2.5.13.}$$

where b and t_0 are constants and together with t_L are called shape, scale, and location parameters, respectively. Once formula 2.5.13 is established, the shape parameter, b, is not necessarily less than unity. This distribution function has no theoretical basis; its validity should be checked by experience.

The same procedure can be followed in deriving the cumulative failure distribution function with respect to voltage or electric field. This leads to

$$F(V) = 1 - \exp\left\{ - \left(\frac{V - V_L}{V_0}\right)^{1/a} \right\} \tag{2.5.14.}$$

Equations 2.5.13. and 2.5.14. are both called the Weibull distribution after Weibull who first derived these expressions.

It is possible to include the two variables, t and V, and to obtain the two-variable Weibull distribution function which can be written:

$$F(V, t) = 1 - \exp\left\{ - \left(\frac{V - V_L}{V_0}\right)^{1/a} \left(\frac{t}{t_0}\right)^{1/b} \right\} \tag{2.5.15.}$$

Confirmation that this expression satisfies Equation 2.5.10. is straightforward. Examination shows an interesting relation between the two variables; for the same cumulative failure rate

$$(V - V_L)^n t = \text{constant} \tag{2.5.16.}$$

where n = b/a.

Furthermore, the three-variable Weibull distribution function, which includes the volume, v, of the specimen, can be deduced on the basis of Formula 2.5.11. as follows:

$$F(V, t, v) = 1 - \exp\left\{ - \left(\frac{V - V_L}{V_0}\right)^{1/a} \left(\frac{t}{t_0}\right)^{1/b} \left(\frac{v}{v_0}\right) \right\} \tag{2.5.17.}$$

The volume can be replaced by either the electrode area or the cable length.

When an impulse voltage rather than AC or DC voltage is impressed, the repetition number, N, is more appropriate as a representative quantity than the time, t. Thus, instead of Equation 2.5.15., we obtain

$$F(V, N) = 1 - \exp\left\{ - \left(\frac{V - V_L}{V_0}\right)^{1/a} N^{1/b} \right\} \tag{2.5.18.}$$

and

$$(V - V_L)^n N = \text{Constant} \tag{2.5.19.}$$

B. Characteristics of the Weibull Distribution Function

For convenience, consider the one-dimensional Weibull distribution function (Equation 2.5.12.), assuming that all other parameters are constant. The instantaneous failure rate and the failure probability density function are then given by

$$\lambda(t) = \frac{1}{bt_0^{1/b}} \, t^{(1/b-1)} \qquad (2.5.20.)$$

and

$$f(t) = \frac{1}{bt_0^{1/b}} \, t^{(1/b-1)} \exp\left\{ -\left(\frac{t}{t_0}\right)^{1/b} \right\} \qquad (2.5.21.)$$

The two functions $\lambda(t)$ and $f(t)$ are represented schematically in Figure 2.5.2. A clear distinction may be found according to the value of the shape parameter b, unity being the critical value for b.

● b < 1 (Wear-out failure)

The failure rate increases with t. As the value of b decreases, failure tends to occur dramatically at a certain time. When a specimen suffers from deterioration which leads to its final breakdown, it may belong to this type of failure. For this reason it is called the wear-out type of failure. The failure probability function closely approaches the normal or Gaussian distribution function when b = 0.31.

● b = 1 (Random failure)

The instantaneous failure rate is independent of t, so that a specimen is subject to random failure without deterioration. The failure probability function is clearly identical with the exponential distribution function.

● b > 1 (Initial failure)

The instantaneous failure rate decreases with t. Failure events are concentrated in the early stages and are classified as initial failures. Poor manufacture of goods is responsible for this.

The reciprocal of the term b is alternatively referred to as the shape parameter, and often represented by m (m = 1/b). When random to wear-out failures are under consideration, i.e., $b \le 1$, the scatter in data decreases as the parameter, b, decreases. Accordingly, the shape parameter can be an index to data scatter in such cases. Median life, mean life, and the dispersion as determined by the Weibull distribution are

$$t_{median} = t_0 (\ell n \, 2)^b \qquad (2.5.22.)$$

$$t_{mean} = t_0 \, \Gamma(1 + b) \qquad (2.5.23.)$$

$$\sigma^2 = t_0^2 \left[\Gamma(1 + 2b) - \Gamma^2(1 + b) \right] \qquad (2.5.24.)$$

respectively. The term t_0 is called the scale parameter or the characteristic life. Values of $(\ell n \, \alpha)^b$ and $\Gamma(1 + b)$ approach $0.8 \sim 1$ when b is less than unity. Formula 2.5.13. can be transformed to yield

$$\ell n \ell n \, \frac{1}{1 - F(t)} = \frac{1}{b} \, \ell n t - \frac{1}{b} \, \ell n t_0 \qquad (2.5.25.)$$

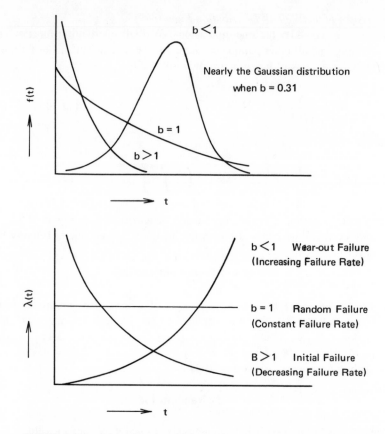

FIGURE 2.5.2.   Schematic representation of f(t) and λ(t) according to the value of the shape parameter b.

A chart of this representation is commercially available and called a Weibull distribution chart. The ordinate of the chart has a linear scale of the left-hand term, and expresses the percentage of the function F(t). The abscissa of the chart expresses the common logarithm of t. The chart can be used conveniently to obtain the shape parameter, the scale parameter, and the median life from experimental results.

All that has been stated regarding t is valid for an N-representative Weibull distribution function. Also, an exactly similar treatment can be applied to a V-representative Weibull distribution function, except for the inclusion of a term V_L (location parameter).

Different features appear with respect to V_L. Nonzero V_L in the Weibull distribution function 2.5.15. means that if the applied voltage is less than the critical or lowest voltage V_L, failure should never occur. It is not always certain whether V_L is finite or zero, it depends on the experimental object of our interest. In an electrical insulation system, for example, V_L might well represent the corona inception voltage. Table 2.5.1. shows some experimental values for various parameters.

2.5.2. Effects of Volume, Area, and Length

Formula 2.5.17. gives the effect of the volume of a specimen in the Weibull distribution function. As noted, this turns out to be the exponential distribution function with respect to the volume. Again as noted, v is not necessarily the volume, but can be the electrode area or the length of a cable with constant diameter. The volume effect may be substantiated by the influence of area and thickness on breakdown voltage. When the volume effect is included, the median life or 50% failure time is given by

Table 2.5.1.
MEASURED VALUES OF N, A, B, AND B/A

Specimen	Experimental condition	n	a	b	b/a
Polyethylene	No void discharge	11	0.18	1.98	11
	With void discharge	$3.7 \sim 11$	—	$0.7 \sim 0.25$	
Polystylene	No discharge	9	0.13	1.21	9.3
Plastic cable	—	14.7	0.17	2.94	17
SF_6 gas	—	—	0.13	—	—
Insulating oil	—	$17.6 \sim 19.2$	$0.05 \sim 0.12$	—	—
Oil-impregnated paper	—	—	0.04	—	—
XLPE cable	Impulse voltage	$9.6 \sim 13$	$0.06 \sim 0.07$	$0.74 \sim 1.62$	—

$$t_m = t_0 (\ell n2)^b \left(\frac{V - V_L}{V_0} \right)^{-b/a} \left(\frac{v}{v_0} \right)^{-b} \qquad (2.5.26.)$$

Alternatively, the 50% failure voltage is

$$V_m = V_0 (\ell n2)^a \left(\frac{t}{t_0} \right)^{-a/b} \left(\frac{v}{v_0} \right)^{-a} \qquad (2.5.27.)$$

These equations infer that, in designing a cable, insulating materials or insulation structures should have values of a and b as small as possible, and values of n ($= b/a$) as large as possible. Having once established the above two formulas experimentally, the life expectancy of a long cable can be estimated from data obtained for short cables, assuming that all other parameters remain the same. This is very useful.

2.5.3. Mixed Weibull Distribution

The Weibull distribution stated above provides a linear relationship between lnln $\{1/(1 - F(v,t))\}$ and ln t or $\ln(V - V_L)$. However, it is often found that the data exhibit a curve rather than a straight line in Weibull plot. Nevertheless, effort has concentrated on trying to modify the distribution rather than abandon it.

Some experimental curved characteristics of Weibull plots have been interpreted by mixing two Weibull distribution functions with different shape parameters, as schematically shown in Figure 2.5.3. It is seen in this figure that two cases are available. One is convex (Case 1), while the other is concave (Case 2). The two cases can be assigned to two different failure modes.

A. Additively Mixed Distributions

If the two modes are serial and mutually incompatible, then the distribution function takes the form

$$F_{add}(V, t) = p_1 F_1 (V, t) + p_2 F_2 (V, t) \qquad (2.5.28.)$$

$p_1 + p_2 = 1$ and one obtains the additively mixed Weibull distribution derived from Formula 2.5.15. as follows:

$$F_{add}(V, t) = 1 - \sum_{i=1}^{2} p_i \exp \left\{ - \left(\frac{V - V_{Li}}{V_{0i}} \right)^{1/a}_i \left(\frac{t}{t_{0i}} \right)^{1/b}_i \right\}$$

$$(2.5.29.)$$

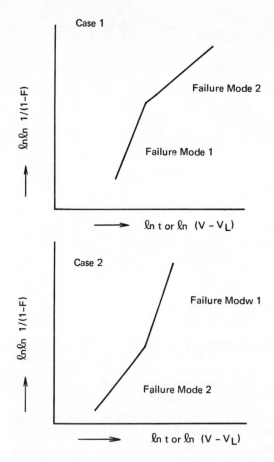

FIGURE 2.5.3. Schematic representation of mixed Weibull distribution function.

By choosing suitable parameters, this expression can interpret the two cases of Figure 2.5.3. However, it should be pointed out that the curve fitting method is meaningful only if the distribution chosen takes account of the prevailing physical and technical conditions and if the plotted data are within the confidence band of the distribution curve.

Concave characteristics for the voltage-Weibull distribution are found for XLPE cables.[199] It is postulated that there are two failure modes, one for breakdown from the middle of the insulation wall and the other for breakdown from the surface of the conductor shielding layer, as shown in Figure 2.5.4. It is believed that the former originates from a corona discharge in a void; this corresponds to failure mode 1 of Case 1 in Figure 2.5.3. The latter, which corresponds to failure mode 2 of Case 1, is possibly caused by tree initiation from a projection on the semiconducting layer. This experimental result indicates that corona discharges leading to breakdown may take place at lower voltages while treeing may be initiated at higher voltages.

Such a statistical failure profile as Case 2 of Figure 2.5.3. can be explained in terms of the additively mixed distribution. It is then required that failure mode 1 be incompatible with failure mode 2. In this case the cable insulation system should be improved because it is implied that macroscopic defects may be involved.

Generally speaking, additively mixed distributions are to be expected when the specimens constituting a random sample can be distinguished with regard to their breakdown behavior by, for instance, differences in prior history, method of production, and materials.[200] In

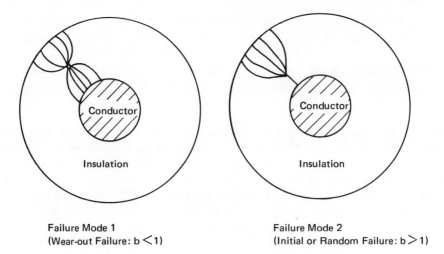

Failure Mode 1
(Wear-out Failure: b <1)

Failure Mode 2
(Initial or Random Failure: b >1)

Modes 1 and 2 are mutually incompatible.

FIGURE 2.5.4. Two speculated failure modes to explain convex characteristics of Weibull plots for XLPE cables.

other words, the profile of the distribution for lot 1 is not always the same as that for lot 2. In Figure 2.5.4., the failure modes are divided into two groups whose breakdown patterns are quite different. This is not a universal requirement, but it is sufficient that the two modes have distinctive shape parameters.

B. Multiplicatively Mixed Distributions

Consider a system consisting of two failure modes which are serial and mutually independent. Then we obtain

$$F_{mul}(V, t) = 1 - \exp\left\{ -\sum_{i=1}^{2} \left(\frac{V - V_L}{V_{oi}}\right)^{1/ai} \left(\frac{t}{t_{oi}}\right)^{1/bi} \right\}$$

$$(2.5.30.)$$

Fischer and Röhl[200] applied the above formula to convex characteristics of Weibull plots (Case 2) for the short-term breakdown strength of polyethylene test specimens and tried to correlate failure mode 2 with breakdowns due to technological influences (impurities etc.), and failure mode 1 with breakdowns attributable to the material itself. Since in each test piece both possibilities of breakdown exist, a division into component lots — compared with the additively mixed distribution — is no longer possible.

For the purpose of reliability and design, Matsuba tried to apply this treatment to a system comprising a cable and its accessories. He proposed that insulation coordination between the two should be determined by the relation[205]

$$\left(\frac{V}{V_{01}}\right)^{1/a_1} \left(\frac{t}{t_{01}}\right)^{1/b_1} \left(\frac{v_1}{v_{01}}\right) = \left(\frac{V}{V_{02}}\right)^{1/a_2} \left(\frac{t}{t_{02}}\right)^{1/b_2} \left(\frac{v_2}{v_{02}}\right)$$

$$\underbrace{\qquad\qquad\qquad}_{\text{cable component}} \qquad \underbrace{\qquad\qquad\qquad}_{\text{accessory component}}$$

$$(2.5.31.)$$

It is also possible to consider a system with two parallel failure modes in which one failure is followed by the other although individual failures in each mode obey the usual Weibull distribution function. This might be the situation when one failure mode is followed by the second, which determines the final breakdown of the specimen. For example, a void discharge can erode the inner surface of a cavity to form a pit from which a tree grows; this in turn leads to total breakdown. The cumulative failure distribution function may be represented by

$$F_p = F_1 \cdot F_2 \qquad (2.5.32.)$$

No investigation has been made of such a parallel system with respect to breakdown modes.

A few further remarks concerning the possibilities of these Weibull representations are in order. By varying the number and the parameters of the component distribution functions and by appropriately choosing their combination (additive or multiplicative mixture), a given series of measurements may be fitted with sufficient accuracy. What is most important in curve fitting is that t-Weibull and V-Weibull be consistent with each other. Requirements for both characteristics should be satisfied. However, one may claim that good fitting alone is not enough, what physical and technical reasons lead to a mixed distribution of the data and what combinations of the partial distributions are therefore to be expected[200] should also be determined.

2.5.4. Modified Weibull Distribution

Consider an experiment in which a number of similar specimens are subjected to voltage. It is not unreasonable to postulate that the failure probability will include additional scatter unrelated to the material itself. In case of void discharges, it may arise from the variation in height of the void, the gas content, the physical condition of their inner surfaces, and so on. The scatter in needle tip radii or in the gas phase between the metal electrode and the insulating material may be responsible in treeing experiments.

There are two schools of thought for developing modified Weibull distributions: they lead to the distributed scale parameter model and the distributed lowest voltage model. Both are based on the belief that the two shape parameters should be independent of time and applied voltage, that is to say it is assumed that the intrinsic process is the same but the extrinsic process is different. Since failure events are mutually incompatible, the failure rate of the system can be represented by the properly weighted sum of the individual distribution functions with different scale parameters.

A. Distributed Scale Parameter Model [207]

The following assumptions are made: Assumption 1 is that life curves obey the relation $V^n t$ = constant. Assumption 2 is that failure rate is determined by Weibull distribution functions with constant shape parameters and different voltage scales. An individual distribution function is then

$$F_i(V_{0i}, V, t) = 1 - \exp\left\{-K \left(\frac{V}{V_{0i}}\right)^{1/a} \left(\frac{t}{t_0}\right)^{1/b}\right\} \qquad (2.5.33.)$$

and the failure distribution function of the whole system is given by

$$F(V,t) = \sum_{i=1}^{i} P_i F_i(V_{0i}, V, t)$$

$$\text{with} \sum_{i=1}^{i} P_i = 1 \qquad (2.5.34.)$$

It is necessary to define the distribution of V_{0i}. For example, one can simply assume that it follows the Gaussian distribution with an average value V_M and a standard deviation $\sigma(\%)$. The value V_M corresponds to the median value if K is chosen so that $F_i = 1/2$ when $(V/V_{0i}) = 1$ and $(t/t_0) = 1$, i.e., $K = \ln 2$.

Use of normal random numbers, generated by a computer, can facilitate the calculation of the cumulative failure rate defined by Formula 2.5.34. This method is useful in comparing computed results with experimental results having the same trial number — the number of random numbers called and the number of specimens used.

Alternatively, we can evaluate the following expression

$$F(V, t) = \frac{1}{C} \int_{V_M}^{\infty} F_i(V_0) \exp\left\{ -\frac{(V_0 - V_M)^2}{2\sigma^2} \right\} dV_0$$

$$\text{with } C = \int_{V_M}^{\infty} \exp\left\{ -\frac{(V_0 - V_m)^2}{2\sigma^2} \right\} dV_0 \qquad (2.5.35.)$$

Formula 2.5.35. corresponds to Formula 2.5.34. when an infinite number of trials are made. This method successfully explained the time-Weibull distributions obtained for breakdown of certain polymers due to void discharges,[164] as shown in Figure 2.4.4. It transpires that V-Weibull characteristics curve upward. Furthermore, it is interesting to note that t-Weibull characteristics also curve upward, although no intentional modification was made with respect to time, t.

B. Random Lowest Voltage Model (Distributed Location Parameter Model)

As mentioned in Section 2.4.1., V-t characteristics are not linear in double-logarithmic plots. The random scale parameter model cannot explain this profile. Introducing the lowest voltage, V_L, helps to explain this qualitatively, as seen already, but provides no insight on the curved t-Weibull characteristics. A new approach has been made to explain both characteristics.[206]

It is important to stick to physical meaning when a distribution is to be modified. We assume that in a given specimen there is a lowest voltage below which no breakdown will take place, even after an infinite time. The lowest voltage, V_L, may differ from specimen to specimen. In other words, the value, V_L, must obey some distribution function which, for simplicity, can be normal. We then obtain

$$F(V, t) = 1 - \sum_i P_i \exp\left\{ -\left(\frac{V - V_{Li}}{V_0}\right)^{1/a} \left(\frac{t}{t_0}\right)^{1/b} \right\}$$

$$\text{with } \sum_i P_i = 1 \qquad (2.5.36.)$$

or alternatively

$$F(V, t) = 1 - \frac{1}{C} \int_0^V \exp\left\{ -\left(\frac{V - V_L}{V_0}\right)^{1/a} \left(\frac{t}{t_0}\right)^{1/b} \right\} \cdot$$

$$\exp\left\{ -\frac{(V_L - V_M)^2}{2\sigma^2} \right\}$$

$$\text{with } C = \int_0^{\infty} \exp\left\{ -\frac{(V_L - V_M)^2}{2\sigma^2} \right\} dV_L \qquad (2.5.37.)$$

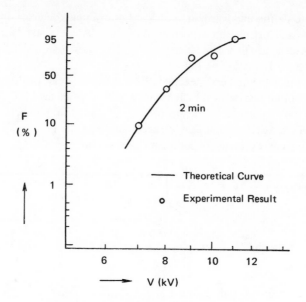

FIGURE 2.5.5.   V-Weibull plots for tree initiation.

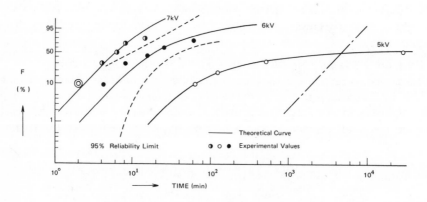

FIGURE 2.5.6.   t-Weibull plots for tree initiation.

When V is smaller than V_L or V_{Li}, we expect no failure and therefore F becomes zero for this condition.

This method has been applied successfully to tree initiation from a needle tip in polyethylene, as is seen from Figure 2.5.5., Figure 2.5.6., and Figure 2.5.7. The scatter in V_L is thought to originate in the varied radii of needle tips which critically define the electric stress.

As pointed out earlier, it is important that both V- and t-Weibull plots be in good agreement as between theoretically predicted results and results obtained experimentally. Modified Weibull distribution functions can no longer be considered to belong to the standard Weibull distribution function, though they may be called thus for convenience.

The analysis of life and reliability of insulation is not yet firmly established, but hopefully it will be developed as a useful tool to investigate insulation behavior. As this occurs it should be correlated with underlying physical processes.

2.6. SPACE CHARGE EFFECTS IN DC CABLES[208-219]

The design of DC cable insulation is clearly different from that of AC cable insulation because of the electric field distribution. Generally speaking, the electric field is resistively

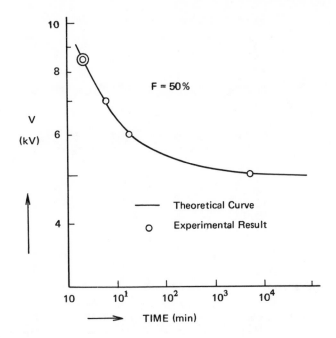

FIGURE 2.5.7.   V-t characteristics for tree initiation.

graded in the former case, while it is capacitively graded in the latter. The electrical resistance of insulation depends appreciably on temperature; it falls as the temperature rises. The capacitance, on the other hand, is more or less constant over the temperature range of interest. As a consequence, the maximum electric stress in the DC case may appear near the insulation shield rather than the conductor shield at full load because the insulation near the conductor is subjected to a higher temperature than that near the outer shield. The result is a change in electric stress partition.

The problem is more complicated when space charge is present, for such charge will modify the electric stress distribution further. If the space charge is insufficiently mobile to follow the change of external voltage, i.e., polarity reversals or intruding voltage transients, breakdown may result in the worst cases.

2.6.1. Homocharge and Heterocharge

Insulation forms the space charge within itself when subjected to a DC voltage. This has been confirmed by many kinds of experiments. "Electrets", in which space charge is pseudo-permanent, are a manifestation of this phenomenon. In certain cases, the charge is stored in what is called a trapping center, which may be analyzed by a technique of thermally stimulated currents (TSC). Charge is either negative or positive.

It is necessary to define which charge, homocharge or heterocharge, occupies the space near the electrode, since this determines the profile of the electric field distribution. Homocharge has the same polarity as the electrode voltage, while heterocharge has the opposite polarity.

Consider a dielectric between two electrodes with which it makes electrically blocking contact; that is to say, no charge carriers are injected into the dielectric from either electrode when an external voltage is applied. Figure 2.6.1.(A) shows this situation. Negative-charge carriers will move toward the anode leaving positive charge behind. Positive-charge carriers will exhibit similar but opposite movement if they are mobile. In either case, positive and negative charge layers are formed near the cathode and the anode, respectively, i.e., "heterocharge" is formed.

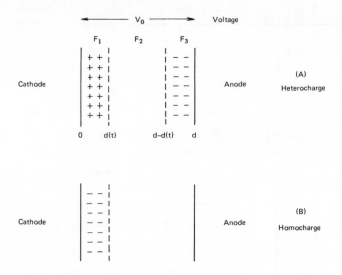

FIGURE 2.6.1.   Heterocharge and homocharge formed near electrodes.

As is clearly seen, the local electric stress near the electrode is much higher than the average electric stress. The local field depends on the concentration of inherent charge carriers or voltage-induced charge carriers. This process is probable in ion-rich materials such as oil-impregnated paper and even in polymers containing ion species.

Consider now a dielectric-metal sandwich system in which electrons can be injected into the dielectric and dominate the charge transfer process within it. Only the behavior of electrons will be described, but the same argument can be reasonably applied to positive holes. Electrons may be emitted from the cathode into the dielectric by either Schottky emission or field emission, (homocharge formation) as schematically shown in Figure 2.6.1.(B). This process is plausible in polymers such as polyethylene and cross-linked polyethylene having little in the way of additives or fillers.

The local electric stress in the homocharge region is lower than the average electric stress. Homocharge appears to be more favorable than the heterocharge; because of this reduction in the local high electric stress a change in voltage polarity may well give rise to electrical instability within the insulation. This is a principal concern in the development of plastic DC cables.

A. Ionic Heterocharge

Consider the system shown in Figure 2.6.1.(A). The electric field at the cathode (or anode), is given by

$$F_c = (\rho_0/\epsilon)d_\infty = (\rho_0 V_0/\epsilon)^{1/2}$$

$$\text{with } d_\infty = (\epsilon V_0/\rho_0)^{1/2} \tag{2.6.1.}$$

where ρ_0 = the density of the space charge, ϵ = the permittivity of the material of interest, d_∞ = the width of the space charge at infinite time. It is assumed that the charge density is constant within the two space charge layers. Table 2.6.1. shows the calculated electric stresses at the cathode when a 250-μm thick oil-impregnated paper is subjected to a voltage of 2500 V, corresponding to an average electric stress of 10 kV/mm (or 250 V/mil). Table 2.6.2. shows the variation of the cathode electric stress and the heterocharge length with applied voltage. Both tables demonstrate that in certain cases the cathode electric field may become significantly higher than the average field, which may possibly cause electron

Table 2.6.1.
CATHODE ELECTRIC STRESSES AND
HETERO-SPACE-CHARGE LENGTHS
DEPENDING ON CARRIER DENSITY

n (cm⁻³)	d_∞ (μm)	Fc(kV/mm)	Fc/Fav	Remarks
10^8	1440	10	1	$d_\infty > d$
10^9	445	10	1	
10^{10}	144	~10	~1	$d_\infty \lesssim d$
10^{11}	44.5	~11.0	~1	
10^{12}	14.4	~34.7	~3	$d_\infty \ll d$
10^{13}	4.45	110	11	
10^{14}	1.44	347	35	
10^{15}	0.45	1110	110	
10^{16}	0.14	3470	347	

Note: Average electric stress: 2.5 kV/250 μm, 10 kV/mm (or 250 V/mil), d = 250 μm or 10 mil.

Table 2.6.2.
CATHODE ELECTRIC STRESSES AND
HETERO-SPACE-CHARGE LENGTHS AS A
FUNCTION OF APPLIED VOLTAGE

V (kV)	F_{av} (kV/mm)	d_∞ (μm)	Fc (kV/mm)	Fc/Fav
0.25	1	4.55	110	110
0.50	2	3.22	145	77.5
1.00	4	2.28	220	55.0
2.50	10	1.44	347	34.7
5.00	20	1.02	490	24.5
10.00	40	0.72	694	17.4
25.00	100	0.46	1097	11.0

Note: n = 10^{14} cm⁻³, d = 250 μm (specimen thickness).

injection from the cathode to neutralize the heterocharge. It is experimentally evident that electrolytic processes are involved in some cases. If only ionic species should participate in the electrical conduction, the current density would be given by

$$J = \frac{\rho_0 \mu V}{d} \left(1 - \frac{2d_\infty}{d} \tanh \frac{t}{2\tau} \right) \text{sech}^2 \left(\frac{t}{2\tau} \right)$$

$$\text{with } \tau = \frac{d_\infty d}{2\mu V} \text{ and } d(t) = d_\infty \tanh \frac{t}{2\tau} \qquad (2.6.2.)$$

where μ = the carrier mobility. As expected, no steady state current would be available, which contradicts experimentally observed phenomena of temporal change in currents. It is therefore logical to postulate that electrons would be injected from the electrode to provide the steady state current.

Actually, the electrical conductivity is dependent on the average electric stress under high stress conditions. Schottky emission is the most likely charge injection process; it can be represented by

$$\sigma = \sigma_0 \exp(\alpha\sqrt{F}) \qquad (2.6.3.)$$

or

$$J = J_0 \exp(\alpha\sqrt{F}) \qquad (2.6.4.)$$

where σ_0 and J_0 are the zero-field conductivity and current density, respectively. This is evident from experiments at comparatively high (average) electric stress on the order of 10 kV/mm to 100 kV/mm.[216,217] Experimentally obtained values of α were found to be in good agreement with the theoretical value. On the other hand, conductivity is determined at lower electric fields typically between 0.2 kV/mm and 0.5 kV/mm, by

$$\sigma = \sigma_0 \exp(\beta F) \qquad (2.6.5.)$$

and is ascribed to the bulk property rather than the electrode effect.[2,15] Formula 2.6.5. suggests that charge carriers are hopping in the array of Coulombic barriers, which would be electrostatically reduced by the applied electric field. It can be interpreted in terms of electrode effects also, assuming electrons are injected through the Coulombic electrode barrier and are free to move afterwards.

B. Electronic Homocharge

When a dielectric is in contact with a metal, charge flow will take place so as to equalize Fermi levels of the two materials and will therefore give rise to either a negatively or a positively charged barrier according to the difference between their work functions, as depicted in Figure 2.6.2. The barrier formed as a consequence of the difference between the metal work function and the dielectric electron affinity is generally called the Schottky barrier. This appears explicitly in Figure 2.6.2.(B) and is positively charged. This well known Schottky emission can be expected to take place from this barrier. The Schottky barrier is often referred to as the exhaustion layer because all the electron donors are exhausted, resulting in the formation of a positive space charge layer. This is likely to happen at the contact between metals and n-type semiconductors, which is sometimes called a rectifying contact because it functions in this way.

Dielectrics with a wide band gap are generally considered to form such a barrier as shown in Figure 2.6.2.(A). In reality, other thin barriers, possibly due to surface states, may be present also, but almost nothing about them is presently understood.

Electrons are injected near the metal electrodes to form negative space charge, which acts as a virtual cathode, supplying as many electrons as required. When voltage is impressed, electrons flow from the virtual cathode, creating an electric current in the dielectric, which is supplemented by other electrons emitted from the metal cathode either thermionicly or by the Schottky mechanism. Current flow is limited by either space charge or emission depending upon the magnitude of the voltage applied.

Homocharge is thus formed. Assuming a uniform distribution of injected charge, we obtain the following expression for the total charge per unit area (from Poisson's equation) and its injection distance:

$$Q_s = \sqrt{n_s \epsilon_i (\phi_i - \phi_m)} \qquad (2.6.6.)$$

$$\xi = \sqrt{\frac{2\epsilon}{n_s e^2}(\phi_i - \phi_m)} \qquad (2.6.7.)$$

where n_s = the density of injected charge carriers and ϵ_i = the permittivity of the dielectric. This negative charge is balanced by a positive charge formed in the metal facing the dielectric, as shown in Figure 2.6.3. The positive charge is screened by electrons in the metal within the Debye shielding length:

ϕ_M: Work Function of Metal

ϕ_I :   Work Function of Dielectric

χ_I :   Electron Affinity of Dielectric

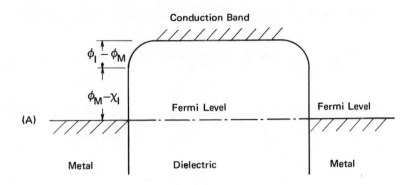

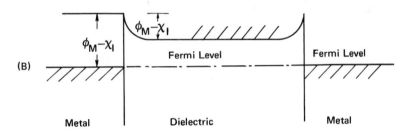

FIGURE 2.6.2.   Two kinds of dielectric-metal contacts — their energy scheme. (A) Electron injection into dielectric. (B) Electron ejection out of dielectric.

$$\xi_m = \sqrt{\frac{\epsilon_m kT}{n_m e^2}} \qquad (2.6.8.)$$

where ϵ_m = the permittivity of the metal and n_m = the density of free electrons in the metal. Since $n_m \gg n_s$, ξ_m is much smaller than ξ.

Suppose now that voltage is applied. Positive charge appears in the anode and is counterbalanced by an equal negative charge appearing in the dielectric. This determines the electric current. Positive charge appearing in the cathode in thermal equilibrium is negligibly small compared to the positive charge formed when voltage is applied. Electric neutrality for the whole system is satisfied when

$$\int_0^d \rho(X)dX = Q_1 + Q_2 \qquad (2.6.9.)$$

$$\int_0^{X_m} \rho(X)dX + \int_{X_m}^d \rho(X)dX = Q_1 + Q_2 \qquad (2.6.10.)$$

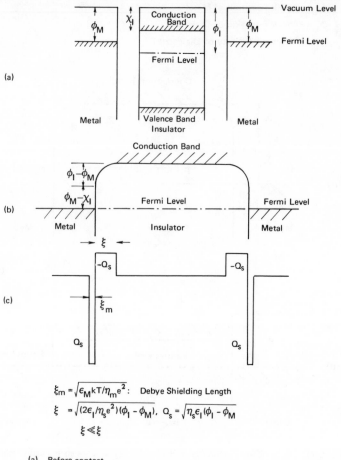

$$\xi_m = \sqrt{\epsilon_M kT/\eta_m e^2}: \quad \text{Debye Shielding Length}$$

$$\xi = \sqrt{(2\epsilon_I/\eta_s e^2)(\phi_I - \phi_M)}, \quad Q_s = \sqrt{\eta_s \epsilon_I (\phi_I - \phi_M)}$$

$$\xi \ll \xi$$

(a)   Before contact

(b)   After contact

(c)   Charge distribution after contact

FIGURE 2.6.3.   Energy diagram for metal-insulator contact.

where Q_1 = the positive charge in the cathode, Q_2 = the positive charge in the anode, and $\rho(X)$ = the density of the negative space charge. The distance X_m can be defined so as to fulfill the following expressions:

$$\int_0^{X_m} \rho(X)dX = Q_1 \text{ and } \int_{X_m}^d \rho(X)dX = Q_2 \qquad (2.6.11)$$

which may be interpreted as saying that the positive charge at either contact is neutralized by an equal amount of negative charge contained between the contact and the plane $X = X_m$. This situation is shown in Figure 2.6.4.(A). It will be noted that X_m corresponds to the distance at which the potential barrier is a maximum.

The cathode in the system can supply as many electrons as required by the applied voltage and consequently the current flow is really space-charged limited. This being the case, the electric field is given by

$$F(X) = \frac{3}{2} \frac{V}{d} \sqrt{\frac{X}{d}} \qquad (2.6.12.)$$

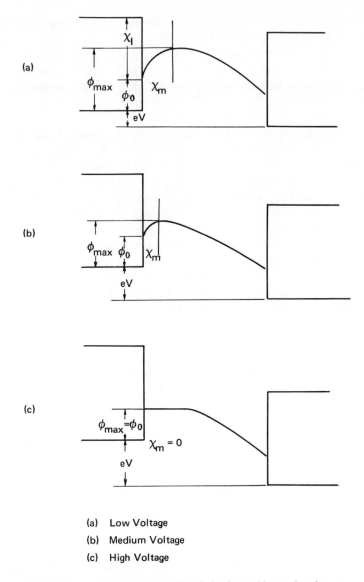

(a)   Low Voltage

(b)   Medium Voltage

(c)   High Voltage

FIGURE 2.6.4.   Potential barrier and applied voltage with regard to electron injection.

is the current flows primarily by diffusion in the cathode region $X < X_m$, and primarily by drift in the anode region $X > X_m$. The derivation of Equation 2.6.12. requires that $F(O) = 0$ to a good approximation. If a single shallow trap level is present, the current is determined by

$$J = \frac{9}{8} \theta \epsilon \mu \frac{V^2}{d^3}$$

(2.6.13.)

where θ = the trapping parameter represented by the ratio of the density of conducting electrons to the density of trapped electrons, ϵ = the permittivity of the dielectric, μ = the electron mobility, d = the interelectrode distance, and V = the applied voltage.

As the applied voltage is increased, more and more electrons are supplied from the virtual cathode or the electron reservoir. This is accomplished by lowering the potential peak ϕ_{max}, thereby facilitating space charge-limited current (SCLC) flow. This process can continue until $\phi_{max} = \phi_0$ at $X_m = 0$. Figure 2.6.4.(B) shows how ϕ_{max} decreases and X_m moves toward the cathode at an intermediate voltage. When the reservoir is exhausted, i.e., $Q_1 = 0$ and $X_m = 0$, ϕ_{max} can no longer be reduced in this manner; the current will be no longer space charge-limited but rather emission-limited. The charge distribution at the onset of emission-limited current will consist of a positive charge on the anode only and negative space charge in the dielectric. Consequently, the potential barrier will decrease monotonically from ϕ_0 at $X = 0$, as shown in Figure 2.6.4.(C).

As voltage is increased further, the barrier can only be lowered by the interaction of the image force with the applied field, i.e., by the Schottky effect. Some negative charge will appear on the cathode, reducing the charge density in the dielectric and limiting the current to a value which can be supplied by the contact. The electric field is determined by two pairs of charges. One charge distribution comprises a negative cathode charge plus an equal positive charge on the anode. It gives rise to a uniform field F_0 directed towards the cathode, where F_0 depends on voltage only. The other charge distribution consists of the negative charge in the insulator plus an equal positive anode charge. It creates a field $F_1(X)$, which is a function of both voltage and position.

The electric field is therefore perturbed by the space charge at any intermediate voltage. At higher voltage it transpires that it is spatially uniform, i.e., $F_1(X) << F_0 = V/d$. Under these conditions, the current density can be written

$$J = J_0 \exp(\alpha\sqrt{F_0}) \tag{2.6.14.}$$

2.6.2. Electric Stress Under Uniform Temperature Distribution Conditions
A. General Method of Deriving the Electric Field Distribution in the Homocharge Case

If a single shallow trap level is present, Poisson's equation is given

$$\operatorname{div}\vec{F} = \frac{en_t}{\epsilon}(1 + \theta) \tag{2.6.15.}$$

where F = the electric field vector, n_t = the density of electrons trapped in the shallow trap level, ϵ = the permittivity, and θ = the trapping parameter. The current density vector is given by

$$\vec{J} = n_f e\mu\vec{F} = n_t\theta e\mu\vec{F} = n_t\theta e\mu\vec{F} \tag{2.6.16.}$$

where n_f = the density of conducting electrons, μ = the electron mobility, and $\theta = n_f/n_t$. Eliminating n_t from the above two equations, we obtain

$$\vec{F}\operatorname{div}\vec{F} = \frac{\vec{J}}{\epsilon\mu}\left(1 + \frac{1}{\theta}\right) = \frac{n_0 e}{\epsilon}\left(1 + \frac{1}{\theta}\right)\vec{F}^*$$

$$\text{with } \vec{F}^* = \vec{J}/\sigma = \vec{J}/n_0 e\mu \tag{2.6.17.}$$

where σ_0 = the zero-field conductivity and n_0 = the density of conducting electrons at thermal equilibrium, i.e., under no electric field. On the other hand

$$\operatorname{div}\vec{J} = 0 \tag{2.6.18.}$$

Since σ_0 is a scalar independent of position, we obtain

$$\operatorname{div}\vec{F}^* = 0 \tag{2.6.19.}$$

It is possible to postulate the potential V* with respect to F* which is determined by

$$\vec{F}^* = - \text{grad } V^* \qquad (2.6.20.)$$

and therefore

$$\nabla^2 \nabla^* = 0 \qquad (2.6.21.)$$

The electric field distribution can be obtained either directly by solving Equation 2.6.17. or indirectly by first obtaining $\vec{F}^*$ from Equations 2.6.20. and 2.6.21. and then substituting $\vec{F}^*$ in Equation 2.6.17.

B. Electric Field in the Plane-Parallel Geometry (Homocharge)

Consider a plane-parallel electrode system for example. Putting the X-axis as the direction of the carrier transport, we obtain

$$\frac{dF^*}{dX} = 0 \qquad (2.6.22.)$$

since no current flows and therefore no electric field is present in a Y-Z plane. Equation 2.6.22. means that F* is independent of both the distance X, and the current continuity. We thus obtain

$$F \frac{dF}{dX} = \frac{n_0 e}{\epsilon} \left(1 + \frac{1}{\theta} \right) F^* \qquad (2.6.23.)$$

This is exactly the same as was obtained from Equation 2.6.17. by Lampert.[108] Since θ is usually much less than unity, this equation leads to

$$F^2(X) = \frac{2J}{\epsilon\mu\theta} X + F^2(0) \qquad (2.6.24)$$

$$V = \int_0^d F(X)dX \qquad (2.6.25.)$$

Now it is important to define one boundary condition, e.g., the electric field at the cathode. If F(0) = 0, as assumed by Lampert,[208] then Formula 2.6.12. holds, which means the flow of SCLC. As voltage is increased further, the current becomes emission-limited, whereupon the cathode electric field is no longer zero but is defined to a close approximation by the Schottky current

$$J = AT^2 \exp \left(- \frac{\phi_M - \chi_I}{kT} \right) \exp \left(\frac{\alpha}{2} \sqrt{F(0)} \right) \qquad (2.6.26)$$

where A is theoretically $(4\pi emk^2/h^3)$ or 1.8×10^6 A/m²deg² but experimentally about 1/3 of this value, and α is $(e^3/\pi\epsilon)^{1/2}$. When the condition $(2Jd/\epsilon\mu\theta)^{1/2} < F(0)$ is satisfied, the electric field is almost uniform. This produces an interesting characteristic; as voltage is increased the electric field distribution tends to be more uniform, although the current is no longer ohmic but superlinearly dependent on voltage.

The Poole-Frenkel effect is likely to be involved in the electrical conduction of such polymers as polyethylene at high electric field.[218,219] Consider shallow traps and deep donors as shown in Figure 2.6.5. and assume that the Poole-Frenkel effect takes place in the deep donors. Then the following equations hold.

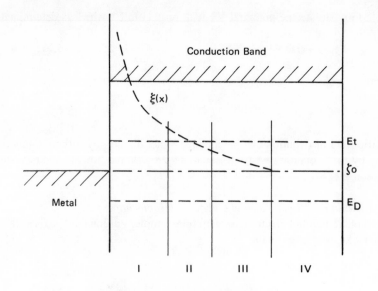

Et : the energy level of shallow traps
E_D : the energy level of deep donors
ζ_0 : the Fermi level in thermal equilibrium
$\zeta(x)$: the Fermi level under voltage application
I : electron injection region
II : trapping region
III : SCLC plus Poole-Frenkel region
IV : Poole-Frenkel region

FIGURE 2.6.5. Energy diagram for a dielectric containing shallow traps and deep donors.

Poisson's equation:

$$\frac{\epsilon}{e} \frac{dF}{dX} = (n(X) - n_0) + (n_t(X) - n_{t_0}) - (p_d(X) - p_{do})$$

(2.6.27.)

current continuity

$$J = n(X)e\mu F(X): \text{ independent of } X \qquad (2.6.28.)$$

density of electrons in the trap levels

$$n(X) = N_t \exp\left[-(E_t - \zeta)kT\right] \qquad (2.6.29.)$$

density of positive holes in the deep donors

$$P_d(X) = N_d \exp\left[-(\zeta - E_d)/kT\right] \qquad (2.6.30.)$$

carrier balance

$$n_0 + n_{t_0} = p_{do} \qquad (2.6.31.)$$

where ϵ the permittivity, e = the electronic charge, F = the electric stress, X = the distance

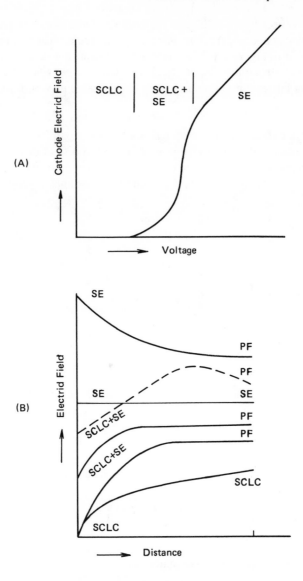

SCLC : Space charge limited current
SE   : Schottky effect
PF   : Poole-Frenkel effect

FIGURE 2.6.6.   Electric fields dependent on voltage and distance
and their relation with associated phenomena.

from the cathode, n_0 = the density of conducting electrons in thermal equilibrium, n_{t0} = the density of trapped electrons in thermal equilibrium, P_{d0} = the density of positive holes in the deep donors in thermal equilibrium, J = the current density, μ = the electron mobility, ζ = the Fermi level, N_c = the effective density of the conduction band, N_t = the density of shallow traps, and N_d = the density of deep donors. We then obtain

$$\frac{dF}{dX} = \left(\frac{en_0}{\epsilon\theta}\right)\ \frac{F^*}{F}\ \left[1 - \left(\frac{F}{F^*}\right)^2 \exp\left(\alpha\sqrt{F}\right)\right] \qquad (2.6.32.)$$

Equation 2.6.31. is not always valid near the cathode. Therefore, it is generally necessary to divide the interelectrode space into three or four regions, each having its respective condition, as shown in Figure 2.6.5. A more detailed treatment is beyond the scope of the book. Profiles of the resulting electric field distributions are schematically presented in Figure 2.6.6. It is interesting to note that even heterocharge may be formed by the Poole-Frenkel effect.

C. Electric Field in Cylindrical Geometry (Homocharge)

We shall now discuss the space charge modified electric field distribution in the cable configuration comprising two concentric cylinders of radii r_c and r_a with insulation between. From Equations 2.6.17. and 2.6.18. in the cylindrical coordinate, we obtain

$$F \frac{d}{dr} (rF) = \frac{rJ}{\epsilon\mu\theta} \equiv F_0^2 = \text{constant} \tag{2.6.33.}$$

and consequently

$$F(r) = F_0 \left(\frac{r_c}{r}\right) \sqrt{\left(\frac{r}{r_c}\right)^2 - 1 + \left(\frac{F_c}{F_0}\right)^2} \tag{2.6.34.}$$

$$V = \int_{r_c}^{r_a} F(r) \, dr \tag{2.6.35.}$$

$F_c = F(r_c)$ corresponds to the condition where the inner cylindrical electrode is negative, i.e., the cathode. The electric stress F_c at the cathode is either zero or finite according to whether the space charge-limited or the emission-limited condition applies. When $F_c = 0$, the field increases monotonically from the cathode. Thus, it is much more like the field in the planar configuration.

On the other hand, when the outer cylindrical electrode is negative, we obtain

$$F(r) = F_0 \left(\frac{r_a}{r}\right) \sqrt{1 - \left(\frac{r}{r_a}\right)^2 + \left(\frac{F_a}{F_0}\right)^2} \tag{2.6.36.}$$

When $F_a = 0$, the field decreases monotonically.

D. Electric Field Around Protrusions

For an electrode system consisting of a semi-infinite hyperboloid needle and a plane, as shown in Figure 2.6.7.(A), the electrostatic field at the tip of the needle when $d > 10r$ is given to a good approximation by

$$F_{max} = \frac{2V}{r\ell n \left(1 + \frac{4d}{r}\right)} \tag{2.6.37.}$$

For the planar electrode system with a protrusion in the form of an ellipsoid of revolution, as shown in Figure 2.6.7.(B), the electrostatic field at the tip of the protrusion is

$$F_{max} = \frac{2Va}{r\ell \left[\frac{1}{2} \ell n(4a/r) - 1\right]} \tag{2.6.38.}$$

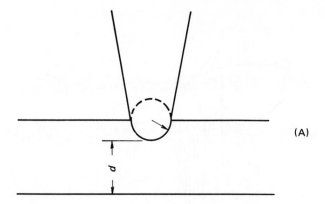

(A)

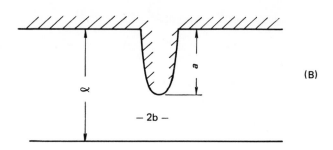

(B)

(A) Hyperboloid of revolution and plane
(B) Ellipsoid of revolution and plane

FIGURE 2.6.7. Electrode systems. (A) Hyperboloid of revolution and plane. (B) Ellipsoid of revolution and plane.

Sometimes a needle-plane electrode system can be approximated by a hyperboloid of revolution. Figure 2.6.8. shows the hyperboloid coordinates. We consider only the Z-axis of the hyperboloid of revolution and use λ instead of $Z(=c\lambda)$, thereby obtaining

$$F(X) \frac{dF(\lambda)}{d\lambda} = \frac{n_0 e}{\epsilon} F^*(\lambda) \qquad (2.6.39.)$$

and therefore

$$F^2(\lambda) - F^2(\lambda_0) = \frac{2n_0 e}{\epsilon} \int_{\lambda}^{\lambda_0} F^*(\lambda)d\lambda$$

$$= \frac{2n_0 e}{\epsilon} [V^*(X) - V^*(\lambda_0)] \qquad (2.6.40.)$$

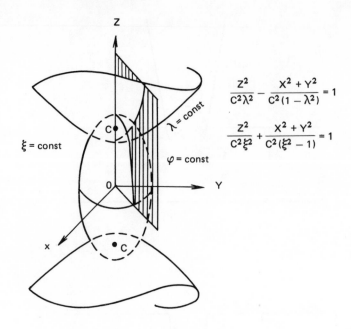

$$\frac{Z^2}{C^2\lambda^2} - \frac{X^2 + Y^2}{C^2(1-\lambda^2)} = 1$$

$$\frac{Z^2}{C^2\xi^2} + \frac{X^2 + Y^2}{C^2(\xi^2 - 1)} = 1$$

λ = const : hyperboloid of revolution
ξ = const : ellipsoid of revolution
φ = const : conjugated axis

FIGURE 2.6.8.   Coordinate of hyperboloid of revolution.

Solutions for $F^*(\lambda)$ and $V^*(\lambda)$ are calculated as follows:

$$F^*(\lambda) = F_0^* \frac{2}{\ln[(1+\lambda_0)/(1-\lambda_0)]} \quad \frac{1}{1-\lambda^2} \tag{2.6.41.}$$

$$V^*(\lambda) = V_0^* \frac{\ln[(1+\lambda)/(1-\lambda)]}{\ln[(1+\lambda_0)/(1-\lambda_0)]}$$

with

$$F_0^* = V_0^*/Z_0 = V_0^*/c\lambda_0 \tag{2.6.42.}$$

In this way we obtain

$$F(\lambda) = F_{so} \left(\eta^2 + \frac{\ln[(1+\lambda)/(1-\lambda)]}{\ln[(1+\lambda_0)/(1-\lambda_0)]} \right)^{1/2}$$

where

$$F_{so} = \sqrt{\frac{2Cn_{to}eV^*}{\epsilon}} \quad \text{and} \quad \eta = F(0)/F_{so} \tag{2.6.43.}$$

Replacing F^* and V^* by F and V leads to the space charge-free solution of the electrostatic field and voltage distribution. Figure 2.6.9. contrasts the electrostatic space charge-free, field distribution (solid line) with the space charge-modified electric field distribution (dotted

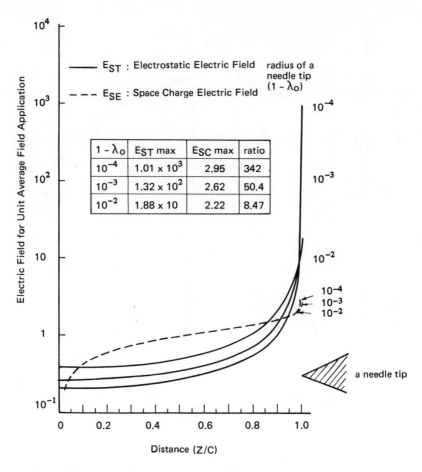

The figure contains the following legend and table:

- E_{ST} : Electrostatic Electric Field
- E_{SE} : Space Charge Electric Field

radius of a needle tip $(1 - \lambda_o)$

$1 - \lambda_o$	E_{ST} max	E_{SC} max	ratio
10^{-4}	1.01×10^3	2.95	342
10^{-3}	1.32×10^2	2.62	50.4
10^{-2}	1.88×10	2.22	8.47

a needle tip

FIGURE 2.6.9. Electrostatic and space charge affected electric field distribution in a needle-plane electrode system.

lines). These curves are obtained when the electric field at the plane electrode is set to zero. It is not easy to formulate solution when the needle is an injecting contact or the cathode satisfying the space charge-limited condition. A solution can be derived from the regional approximation method. The field distribution surely has a peak near the tip of the needle electrode. Profiles of the field distributions are summarized in Figure 2.6.10.

2.6.3. Approximate Electric Stress Formulas for Cable Design

Space charge-modified electric fields have been generally treated in the last section. In addition, as we have partly explained in discussing the Poole-Frenkel effect, the supra-linearity of the flowing current on the electric field strength should be taken into consideration. Mechanisms for the nonlinearity still remain controversial. Furthermore, there is usually a spatial and especially radial variation in temperature in cable insulation in service, which, in turn, causes a variation in electrical conductivity, because conductivity is sensitive to temperature. The general treatment explained earlier could be applied to this situation and should be investigated further. As things stand, a different approach has been explored. Some approximate methods have been developed for cable design, which include the influence of electric field and temperature on the electrical conductivity of cable insulation.

A. Spatially Distributed Conductivity

Consider a cable composed of a cylindrical conductor having a radius R_c, with insulation of thickness d, and a conductive screen having radius $R_s = R_c + d$. Under load conditions,

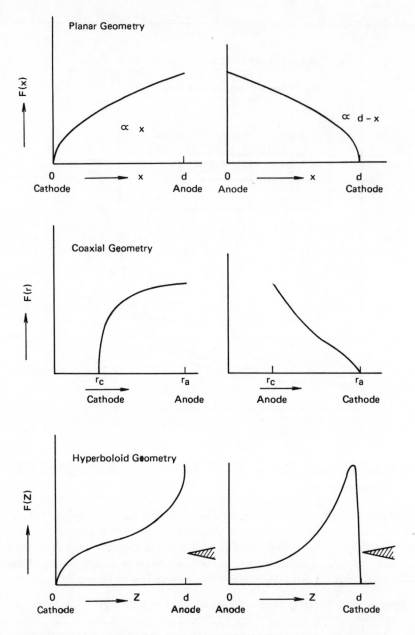

FIGURE 2.6.10. Space charge modified electric field distribution for three different electrode systems.

the conductor experiences joule heating, while the screen is cooled by the ambient. Since the thermal conductivity of the insulation can be assumed constant for the limited temperature difference we are dealing with, the temperature $T(r)$ in the insulation varies with the radius r according to the well known law

$$T(r) = T_s + \frac{T_c - T_s}{\ln(R_s/R_c)} \ln \frac{R_s}{r} \qquad (2.6.44.)$$

where T_c and T_s are the temperatures of the conductor and the screen, respectively. In Equation 2.6.44., the heat generated by dielectric losses is disregarded because it is insignificant under a steady state condition.[215]

a. Conductivity Expression $\sigma = \sigma_0\, e^{\alpha T} F^p$

If the conductivity is expressed in this way, the resulting electric field and the voltage at a distance r from the center of a cable conductor are given by

$$F(r) = \frac{\delta r^{(\delta-1)}}{R_c^\delta - R_s^\delta}\, V \tag{2.6.45.}$$

$$V(r) = \frac{R_c^\delta - r^\delta}{R_c^\delta - R_s^\delta} \tag{}$$

with

$$\delta = \frac{p}{p+1} + \frac{\alpha}{p+\perp}\frac{\Delta T}{\ln(R_s/R_s)} \tag{2.6.46.}$$

where V = the DC applied voltage (kV), R_c = the conductor radius (mm), R_s = the insulation radius (mm), ΔT = the temperature difference between the conductor and the screen (°C), σ = the conductivity (ω/cm) F = the electric field (kV/mm), T = the temperature (°C), α = the temperature coefficient of conductivity (°C^{-1}), and p = the stress coefficient of conductivity. The terms α and p are material constants. It is generally accepted that $\alpha \simeq 0.1$, $p \simeq 1$ for oil-impregnated paper and $\alpha \simeq 0.05$, $p \simeq 1$ for XLPE insulation.

b. Conductivity Expression $\sigma = \sigma_0\, e^{(\alpha T + \beta F)}$

From Equation 2.6.44. and the above expression, we obtain

$$F(r)\, e^{\beta F(r)} = (r/R_c)^{\gamma-1}\, F(r_a)\, e^{\beta F(R_i)} \tag{2.6.47.}$$

with

$$\gamma = \frac{\alpha \Delta T}{\ln(R_i/R_c)}$$

where $F(r_a)$ = the average electric field $V/(R_i - R_c)$ (kV/mm), r_a = the distance (mm) at which the average electric field is realized, and β = the stress coefficient of conductivity (mm/kV). Typical values for β are 0.03 and 0.15 for oil-impregnated paper and XLPE insulation, respectively. The electric field can be computed using Equation 2.6.47. together with

$$I_0 = 2\pi r\sigma F \tag{2.6.48.}$$

and

$$V_0 = \int_{R_0}^{R_s} F(r)\,dr \tag{2.6.49.}$$

Figure 2.6.11. shows some examples of computed electric field distributions. It is interesting to compare the above expressions for the electric field distribution with the well known expression in space charge-free and isothermal conditions which is

$$F(r) = \frac{V_0}{r\ln(R_s/R_c)} \tag{2.6.50.}$$

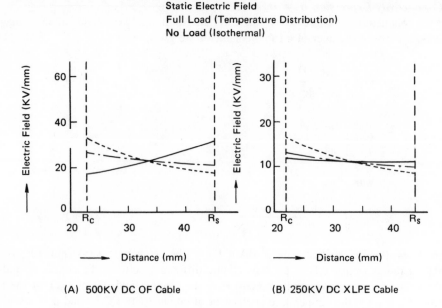

Static Electric Field
Full Load (Temperature Distribution)
No Load (Isothermal)

(A) 500KV DC OF Cable   (B) 250KV DC XLPE Cable

FIGURE 2.6.11.   Calculated electric field distributions.

The formulas obtained above are considered approximate because they do not include any space charge effect. Since it is known that space charge is involved in determining the electric field in insulation, more rigorous formulas should be established, especially for plastic cable insulation which is susceptible to electron injection from the electrodes.

Further problems are associated with the temperature and field dependence of the conductivity. Carrier transport is thermally activated and therefore should be expressed in terms of $\exp(-W/kT)$, where W, k, and T are the apparent activation energy for carrier transport, Boltzmann's constant, and the temperature, respectively. There are two expressions available for the high field conductivity: $\exp(\alpha\sqrt{F})$ and $\exp(\beta F)$. In other words:

$$\sigma = \sigma_0 \exp\left(-\frac{W - \alpha_0 \cdot \sqrt{F}}{kT}\right) \tag{2.6.51.}$$

$$\sigma = \sigma_0 \exp\left(-\frac{W - \beta_0 \cdot F}{kT}\right) \tag{2.6.52.}$$

Comprehensive methods should be explored which correctly take into account the effect of space charge, the temperature dependence, and the high field effect.

2.6.4. Polarity Reversal and Transient Voltage Superposition

The problem of transients is important in DC cable insulation. When DC voltage is applied to insulation, formation of space charge begins, either by charge injection or by charge flow as described in Section 2.6.1.; it reaches its steady state distribution after a certain period of time. If such space charge, when it forms, cannot move in response to a sudden change in voltage, there is a danger of insulation failure. Such transients in DC cables are caused by reversal of polarity, by the malfunction of a converter, or by the intrusion of lightning surges. As is well known, voltage is reversed in order to reverse power flow.

It is claimed that the concept of spatial resistivity (or conductivity) distribution is sound for estimating the electric field distribution in DC cable insulation, but unfortunately it is invalid for investigating the behavior of the electric field when a voltage change occurs.

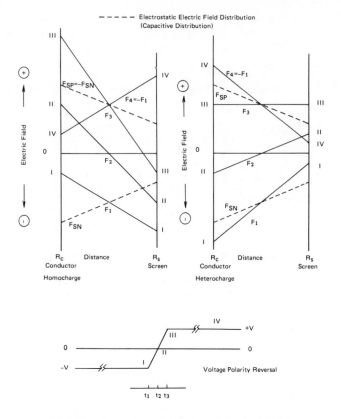

FIGURE 2.6.12.   Successive redistribution of electric field in response to voltage change (schematic interpretation).

Charge persists and it is not released immediately after the removal of voltage. Figure 2.6.12. shows a sequence of electric field redistributions according with voltage changes for both homocharge and heterocharge cases.

Now consider for simplicity only the effect of space charge. After voltage ($-V$) is applied for a sufficiently long time, the steady state field distribution F_1 is reached. It deviates by the differences ($F_1 - F_{sn}$) from the space charge-free electric field distribution F_{sn}. When the voltage starts to change at t = t_1 and is reduced to zero at t = t_2, only the field difference ($F_1 - F_{sn}$) remains, this is represented by F_2 in Figure 2.6.12. When the voltage changes to $+V$ at t = t_3, the resulting electric field F_3 is given by the sum of the space charge-free electric field F_{sp} and the remaining space charge electric field F_2. That is

$$F_3 = F_{sp} + F_2 = 2F_{sp} + F_1 = 2F_{sp} - |F_1| \qquad (2.6.53.)$$

After sufficient time, long enough to allow the space charge to redistribute, the electric field reaches F_4, which is essentially equal to F_1 but of opposite polarity. It is clear from Figure 2.6.12. that the transient electric field at the conductor surface may become much higher than the steady state electric field. During the transient, a polarity reversal in a cable under stress can even cause a doubling of the maximum gradient.

Heterocharge formation seems likely in oil-impregnated paper. The insulation is therefore in less jeopardy when a transient polarity change occurs. On the other hand, homocharge is probable in polymers, sufficient to modify the electrostatic field. Persistent homocharge may result in insulation failure.

The electric field in plastic cable insulation, under the action of a DC voltage and with superposed transient voltage, is apparently more complicated than anticipated. In addition

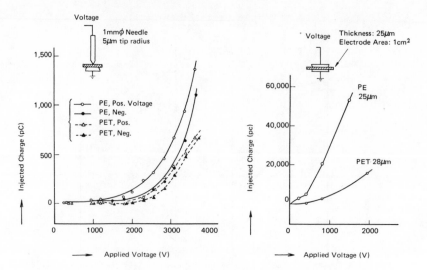

FIGURE 2.6.13. Characteristics of charge injected into two polymers by the application of DC voltage. (From Tanaka, T. and Greenwood, A., *IEEE Trans. Power Appar. Syst.*, 95(5), 1751, 1978. With permission.)

to the complexity of charge injection from electrodes, ionic species existing in polymers may influence the electric field distribution. The variation in temperature distribution due to load change is a further factor.[215]

One technique for improving DC plastic cables is to mitigate the effects of injected homocharge. There are two basic methods: (Method I) suppress the electron injection; (Method II) neutralize the injected electrons. There are several ways to implement these methods:

Method I
1. Trap electrons near an electrode. The γ-ray graft polymerization of polar polymers onto polyethylene may work
2. Use higher work function metals
3. Reduce the work function and electron affinity of the insulation
4. Reduce the density of electron traps
5. Use semiconducting layers to ease the electric field. This approach is theoretically valid provided it does not enhance electron injection

Method II
1. Use materials with higher electron mobility
2. Neutralize electronic charge by positive ions. Impurities can be added to increase ionic conduction

Methods I.2., I.3., I.4., and II.1. were derived from electric field calculation in a plane-parallel electrode system by taking the Schottky effect and the Poole-Frenkel effect into consideration.[211] An investigation based on Method I.3. revealed that the degree of charge injection depends on the materials involved. It indicated the following ascending order of injection level:[212,213]

$$PET < PP < CTA < PE < PC$$

Figure 2.6.13. shows the dependence of injected charge on voltage for polyethylene (PE) and polyethylenetetraphthalate (PET). Low injection materials may be useful for DC insulation.

REFERENCES

1. *Underground Systems Reference Books,* Edison Electric Institute, New York, 1957.
2. *Electrical Transmission and Distribution Reference Book,* Westinghouse Electric Corporation, Pittsburgh, 1964.
3. **Iizuka, K., Ed.,** *Power Cable Technology Handbook,* (in Japanese), Denki-Shoin, Tokyo, 1974.
4. **Arkell, C. A.,** *Underground High Voltage Power Cables,* BICC Cable Division Post Graduate Education Centre, United Kingdom Atomic Energy Association, Harwell, Buckinghamshire, England, 1968.
5. **AEIC No. 4-69,** *Specifications for Impregnated-Paper-Insulated Low-Pressure Oil-Filled Cable,* 6th ed., Association of Edison Illuminating Companies, New York, 1969.
6. **AEIC No. 2-67,** *Specifications for Impregnated-Paper-Insulated Cable- High-Pressure Pipe-Type,* 2nd ed., Association of Edison Illuminating Companies, New York, 1967.
7. **AEIC No. 5-75,** *Specifications for Polyethylene and Cross-Linked Polyethylene Insulated Shielded Power Cables Rated 5 through 69 kV,* 5th ed., Association of Edison Illuminating Companies, New York, 1975.
8. **AEIC G1-68,** *Guide for Application of AEIC Maximum Insulation Temperatures at the Conductor for Impregnated-Paper-Insulated Cables,* 2nd ed., Association of Edison Illuminating Companies, New York, 1968.
9. **Bahder, G., McKean, A. L., and Carrol, J. C.,** Development and installation of 138 kV cable system for tests at EEI Waltz Mill Station: cable 21, *IEEE Trans. Power Appar. Syst.,* 91(4), 1427, 1972.
10. **McAvoy, F. M. and Waldron, R. C.,** EEI-manufactures 500/550 kV cable research project. Cable A-high pressure oil paper pipe types, *IEEE Trans. Power Appar. Syst.,* 90(1), 204, 1971.
11. **Eager, G. S., Jr. and Silver, D. A.,** Development and installation of 138 kV cable for tests at EEI Waltz Mill Station: cable 22, *IEEE Trans. Power Appar. Syst.,* 91(4), 1434, 1972.
12. **Weast, R.,** *Handbook of Chemistry and Physics,* 56th ed., CRC Press, Cleveland, Ohio, 1975—1976.
13. *Rika-Nenpyo (Annual Science Handbook),* Tokyo College of Astronomy, Maruzen Publ., Tokyo, 1977.
14. **Von Hippel, A. R.,** *Dielectric Materials and Applications,* MIT Press, Cambridge, John Wiley & Sons, New York, 1954.
15. *Modern Plastics Encyclopedia 1969—1970,* McGraw-Hill, New York.
16. Electronic Properties Data Center Data Sheets, 1977.
17. **Fukagawa, H.,** *Thermal Resistivity of XLPE Cable Insulation,* (in Japanese), Tech. Rep. No. 74068, Central Research Institute of Electric Power Industry, Tokyo, 1975, 1.
18. **Hansen, D. and Bernier, G. A.,** Thermal conductivity of polyethylene, *Polymer Eng. Sci.,* 12(3), 204, 1972.
19. **Hansen, D. and Ho, C.,** Thermal conductivity of high polymers, *J. Polymer Sci. Pt. A,* 13, 659, 1965.
20. **Mildner, R. C.,** The short circuit rating of thin metal tape cable shields, *IEEE Trans. Power Appar. Syst.,* 87(3), 749, 1968.
21. **Gazzana Priaroggia, P., Occhini, E., and Palmieri, N.,** *A Brief Review of The Theory of Paper Lapping of A Single-Core High-Voltage Cable,* IEEE Monograph No. 390 S, Piscataway, N.J., July 1960, 1.
22. **Yamamoto, T., Nakamoto, S., Yamamoto, M., and Take, Y.,** Mica-loaded paper for EHV power cable, *IEEE Trans. Power Appar. Syst.,* 88(6), 890, 1969.
23. **Lyman, T.,** *Metal Handbook,* Vol. 1, American Metals Society Metals Park, Ohio, 1961.
24. **IPCEA P-34-359,** *AC/DC Resistance Ratios at 50 Hz,* Insulated Power Cable Eng. Assoc., Belmont, Mass., 1973.
25. **Eich, E. D.,** EEI-manufacturers 500/55 kV cable research project. Cable B-high pressure oil paper pipe type, *IEEE Trans. Power Appar. Syst.,* 90(1), 212, 1971.
26. **Mckean, A. L., Merrell, E. J., and Moran, J. R., Jr.,** EEI-manufacturers 500/550 kV cable research project. Cable C-high pressure oil paper pipe type, *IEEE Trans. Power Appar. Syst.,* 90(1), 224, 1971.
27. **IPCEA P-45-482,** *Short-Circuit Performance of Metallic Shielding and Sheaths of Insulated Cables,* Insulated Power Cable Eng. Assoc., Belmont, Mass., 1963.
28. **Brookes, A. S.,** The design of specially bonded cable systems, *Electra,* May 1973, p. 55.
29. **CIGRE SC-21 WGO 7-38, Document 07-38,** *The Design of Specially Bonded Cable Circuits (Part II),* CIGRE, Paris, 1974.
30. **Ball, E. H., Occhini, E., and Luoni, G.,** Sheath overvoltages in high-voltage cables resulting from special sheath-bonding connections, *IEEE Trans. Power Appar. Syst.,* 84(10), 974, 1965.
31. **Masio, G. and Occhini, E.,** *Overvoltage on Anti-Corrosion Sheaths of High-Voltage Cables with Particular Reference to Long Submarine Cables,* CIGRE No. 224, CIGRE, Paris, 1964, 1.
32. IEC 287, Clause 6.6., International Electrochemical Commission, Geneva.
33. **Haga, K. and Kusano, T.,** Surge phenomena on the sheaths of cross-bonded 3-phase cable lines, *J. Inst. Electr. Eng. Jpn.,* 79, 1580, 1959.
34. **Watson, W. and Ervin, C. C.,** Surge potentials on underground cable sheath and joint insulation, *IEEE Trans. Power Appar. Syst.,* 82(1), 239, 1963.

35. **Miller, K. W.,** Electrical Engineering, Ph.D. thesis University of Illinois, Urbana, 1929, 52.
36. **Imai, T.,** Integral equation for sheath losses of a single core aluminum sheathed power cable, *J. Inst. Electr. Eng. Jpn.,* (in Japanese), 3, 142, 1964.
37. **Bahder, G. and Eager, G. S.,** Review of Recent and Future Underground Transmission Systems, APPA Eng. and Operations Workshop, Washington, D.C., February 26, 1975.
38. **Okamoto, H., Tanaka, T., Fukagawa, H., and Koyama, H.,** *Investigation Research of Sodium Conductor Cable,* Technical Rep. No. 69062, Central Research Inst. of Electric Power Industry, Tokyo, 1970, 1.
39. **Ruprecht, A. E. and Ware, P. H.,** Evaluation of sodium conductor power cable, *IEEE Trans. Power Appar. Syst.,* 86(4), 401, 1967.
40. **Humphrey, L. H., Hess, R. C., and Addis, G. I.,** Insulated sodium conductors, *IEEE Trans. Power Appar. Syst.,* 86(7), 881, 1967.
41. **Mattysse, I. F. and Scoran, E. M.,** The development of connectors for insulated sodium conductor, *IEEE Trans. Power Appar. Syst.,* 86(7), 883, 1967.
42. **Humphrey, L. E., Addis, G. I., and Hess, R. C.,** Drawing of insulated sodium conductor, *IEEE Trans. Power Appar. Syst.,* 86(7), 900, 1967.
43. **Hus, J.,** Bistable operating temperatures and the circuit rating of sodium conductors, *IEEE Trans. Power Appar. Syst.,* 87(2), 367, 1968.
44. **Steeve, E. J.,** Dig-in tests on sodium conductor cables, *IEEE Trans. Power Appar. Syst.,* 87(9), 1775, 1968.
45. **Gerhard, S.,** *Improved Connectors for Insulated Sodium Conductor,* IEEE Paper No. 68 CP45-PWR, 1968 Winter Power Meeting, IEEE, Piscataway, New Jersey.
46. **Eichhorn, R. M. and Addis, G. I.,** *Irradiated Polyethylene Insulation for Sodium Conductor Cable,* IEEE Paper No. 68 CP61-PWR, 1968 Winter Power Meeting, IEEE, Piscataway, New Jersey.
47. **Kelly, T. H. and Guerre, C. G.,** *A Progress Report on Sodium Conductor Power Cables,* IEEE Paper No. 68 CP62-PWR, 1968 Winter Power Meeting, IEEE, Piscataway, New Jersey.
48. **Garrison, R. L.,** *Field Service Experience with Sodium Conductor Cable,* Special Technical Conference on Underground Distribution 69 CP-PWR, IEEE, Piscataway, New Jersey, 386.
49. **Satty, M.,** *Sodium — Its Manufacture, Properties and Uses,* Reinhold, New York, 1956.
50. Why sodium cable failed, *Electr. World,* November 7, 1966.
51. **Graneau, P.,** 138 kV and 230 kV Sodium Conductor Cables, private communications, 1976.
52. **Tanaka, T., Fukuda, T., and Handa, S.,** *Thermal Degradation of Oil-Impregnated Paper for EHV Cable (I),* (in Japanese), Tech. Rep. No. 68006, Central Research Inst. of Electric Power Industry, Tokyo, 1968, 1.
53. **Tanaka, T. and Fukuda, T.,** *Thermal Degradation of Oil-Impregnated Paper for EHV Cable (II),* (in Japanese), Tech. Rep. No. 70005, Central Research Inst. of Electric Power Industry, Tokyo, 1970, 1.
54. **Saito, Y. and Take, Y.,** *Electrical Insulating Paper,* (in Japanese), Corona Publ., Tokyo, Japan, 1969.
55. **de Vos, J. C. and Vermeer, J.,** *Contribution to the Development of Power Cables for High and Extra High-Voltage with A Composite Insulation of Synthetic High Polymer Materials and Oil,* CIGRE Report No. 207, CIGRE, Paris, 1958.
56. **Vermeer, J., Boone, W., Bussik, J., and Brakel, H.,** *TENAX — A New Low-Loss High Temperature Resistance Synthetic Paper for EHV Cables and Other Electrical Equipment,* 1970 Annual Report, CEIDP, National Academy of Sciences, Washington, D.C., 1971, 56.
57. **Buehler, C. A., Burvee, R. W., Doty, C. T., and Ugro, J. V., Jr.,** *Evaluation of a Polyolefin Paper-Oil Composite as an EHV Cable Insulation,* 1970 Annual Report, CEIDP, National Academy of Sciences, Washington, D.C., 1971, 70.
58. **Fujita, H., Itoh, H., and Ichine, T.,** *Synthetic Paper for EHV Cable Insulation,* (in Japanese), Proc. 4th Symp. EIM-IEEJ, Institute of Electrical Engineers of Japan, Tokyo, 1971, 71.
59. **Fujita, H. and Itoh, H.,** Synthetic polymer papers suitable for use in EHV underground cable insulation, *IEEE Trans. Power Appar. Syst.,* 95(1), 130, 1976.
60. **Matsuura, K., Kubo, H., and Sasazima, Y.,** *Oil Impregnated Paper/Plastic Composite,* (in Japanese), Proc. 5th Symp. EIM-IEEJ, Institute of Electrical Engineers of Japan, Tokyo, 1972, 89.
61. **Matsuura, K., Kubo, H., and Miyazaki, T.,** *Development of Polypropylene Laminated Paper Insulated EHV Power Cables,* IEEE Conf. Rec. 1976 Underground T & D Conf., IEEE, Piscataway, New Jersey, 322.
62. **Edwards, D. R. and Melville, D. R. G.,** *An Assessment of the Potential of EHV Polyethylene/Paper Laminate Insulated Self-Contained Oil-Filled Cables,* IEEE Conf. Rec. 1974 Underground T & D Conf., IEEE, Piscataway, New Jersey, 529.
63. **Isshiki, S. and Nakayama, S.,** *Synthetic Paper for Use of UHV Power Cables,* (in Japanese), Rep. IM-71-9, Institute of Electrical Engineers of Japan, Tokyo, 1971.
64. **Yamanoto, T., Isshiki, S., and Nakayama, S.,** Synthetic paper for extra high voltage cable, *IEEE Trans. Power Appar. Syst.,* 91(6), 2415, 1972.

65. **Kojima, Y., Soda, S., Suda, K., and Kinoshita, S.,** *Development of Oil-Impregnated Synthetic Paper for EHV Cable,* (in Japanese), Proc. 5th Symp. EIM-IEEJ, Institute of Electrical Engineers of Japan, Tokyo, 1972, 93.

66. **Sekiguchi, H.,** *Jpn. Plast.,* 8(5), 1974.

67. Yasui, Japanese Patent Open Sho-48-859.

68. **Sato, M., Toyama, Y., Hayashida, K., and Ando, N.,** *Application of Synthetic Insulating Papers for EHV Oil-Filled Cables,* Proc. 5th Symp. EIM-IEEJ, Institute of Electrical Engineers of Japan, Tokyo, 1972, 97.

69. **Kanemura, K., Ando, A., and Numajiri, F.,** *Electric Properties of FEP-Laminate Paper for 1,000 kV Oil-Filled Cable,* Proc. 9th Symp. EIM-IEEJ, Institute of Electrical Engineers of Japan, Tokyo, 1976, 121.

70. Special Issue — Recent insulating materials, (in Japanese), *J. Inst. Electr. Eng. Jpn.,* 95(5), 1, 1975.

71. **Fukagawa, H., Okamoto, H., Tanaka, T., Nakasa, H., Nagao, T., Akimota, T., Hamamatsu, T., and Kaminosono, H.,** *Investigation Research of Cryogenic Power Cables,* Tech. Rep. No. 71021, Central Research Inst. of Electric Power Industry, Tokyo, 1971, 1.

72. **Scott, R. B.,** *Cryogenic Engineering,* D Van Nostrand, New York, 1959.

73. **Iida, O., Ed.,** *Tables of Physical Constants,* Asakura-Shoten, Tokyo, 1969.

74. Japan Chemical Society, **Ed.,** *Chemical Handbook,* Maruzen, Tokyo, 1966.

75. **Mathes, K. N.,** Dielectric properties of cryogenic liquids, *IEEE Trans. Electr. Insul.,* 2(1), 24, 1967.

76. **Bobe, J. and Perrier, M.,** Propriété des isolants solides aux températures cryogéniques, *Rev. Gen. Electr.,* 77, 605, 1968.

77. **Stone, F. T. and McFee, R.,** Dielectric strength of some common electrical insulators in liquid helium and nitrogen, *Rev. Sci. Instr.,* 32, 1400, 1961.

78. **Miura, T., Iwata, Z., and Furuto, Y.,** *Dielectric Strength of Some Materials at Low Temperature,* Proc. 3rd Symp. EIM-IEEJ, Institute of Electrical Engineers of Japan, Tokyo, 1970, 57.

79. **Jefferies, M. J. and Mathes, K. N.,** *Insulation System for Cryogenic Cable,* IEEE PES Winter Meeting TP44-PWR, IEEE, Piscataway, New Jersey, 1969.

80. **Miranda, F. J. and Gazzana Priaroggia, P.,** Self-contained oil-filled cables — a review of progress, *Proc. Inst. Electr. Eng.,* 123(3), 229, 1976.

81. **Kojima, K., Nishi, M., and Matsuura, K.,** *Dissipation Factors in Oil-Filled Cable Insulation,* (in Japanese), Sumitomo Denki Review, Sumitomo Electric Industry, Inc., Osaka, 1964, 67.

82. **Blodgett, R. B. and Gooding, R. H.,** Parameters affecting the increase in dielectric loss, caused by carbon black paper screens for oil paper dielectrics, *IEEE Trans. Power Appar. Syst.,* 83G, 121, 1964.

83. **Bahder, G. and Garcia, F. G.,** Electrical characteristics and requirements of extruded semiconducting shields in power cables, *IEEE Trans. Power Appar. Syst.,* 90(3), 917, 1971.

84. **Forster, E. O.,** Electrical conduction mechanism in carbon filled polymers, *IEEE Trans. Power Appar. Syst.,* 90(3), 913, 1971.

85. Lion-AKZO Co., Ltd., Pamphlet — Ketjen Black, 1976, 1.

86. **Taniguchi, K.,** private communication.

87. **Havard, D. G.,** *Selection of Cable Sheath Lead Allows for Fatigue Resistance,* IEEE PES Meeting F76 441-6, IEEE, Piscataway, New Jersey, 1, 1976.

88. **Saito, Y., Ed.,** *Handbook of Electrical Insulating Materials,* Nikkan-Kogyo Press, Tokyo, 1974.

89. **Loeb, L. B.,** *Electrical Corona,* University of California Press, Berkeley, 1965.

90. **Loeb, L. B. and Meck, J. M.,** *The Mechanism of the Electric Spark,* Stanford University Press, Palo Alto, 1941.

91. **Meek, J. M. and Craggs, J. D.,** *Electrical Breakdown of Gases,* Clarendon Press, Oxford.

92. **Whitehead, S.,** *Dielectric Breakdown of Solids,* Clarendon Press, Oxford, 1951.

93. **O'Dwyer, J. J.,** *The Theory of Dielectric Breakdown of Solids,* Clarendon Press, Oxford, 1964.

94. **Adamczewski, I.,** *Ionization, Conductivity and Breakdown in Dielectric Liquids,* Taylor & Francis Ltd., London, 1969.

95. **Seitz, F.,** On the theory of electron multiplication in crystals, *Phys. Rev.,* 76, 1376, 1949.

96. **Inuishi, Y.,** Electrical conduction of dielectrics, *J. Inst. Electr. Eng. Jpn.,* 94, 779, 1974.

97. **Kitani, I. and Arii, K.,** Breakdown time-lags and their temperature characteristics of polymer, *Trans. Inst. Elect. Eng. Jpn.,* 94-A, 251, 1974.

98. **Miyairi, K., Sawa, G., and Ieda, M.,** Dielectric breakdown of polyethylene at low temperatures, *J. Inst. Electr. Eng. Jpn.,* 91, 1962, 1971.

99. **Tanaka, T.,** *Breakdown and Degradation of Polyethylene,* Tech. Rep. CRIEPI No. 67027, Central Research Inst. of Electric Power Industry, Tokyo, 1967, 1.

100. **Artbauer, J.,** The intrinsic electric strength of polymers and its relation to the structure. II. Theoretical, *Acta Technica CSAV,* No. 3, 429, 1966.

101. **Yahagi, K. and Maeda, Y.,** *Elongation of γ-ray-Irradiated Polyethylene and Its Electric Strength,* (in Japanese), IEEJ-EIM-76-63, Institute of Electrical Engineers of Japan, Tokyo, 1976, 1.

102. **Saito, Y. and Take, Y.,** *Electrical Insulating Paper,* Corona Publ., Tokyo, 1969.

103. *Handbook for Electrical Discharge,* (in Japanese), revised ed., Institute of Electrical Engineers of Japan, Tokyo, 1974.

104. **Sharbaugh, A. H. and Watson, P. K.,** *Conduction and Breakdown in Liquid Dielectrics Progress in Dielectrics,* Vol. 4, Heywood, London, 1962, 199.

105. **Krauscki, Z.,** Breakdown of liquid dielectrics, *Proc. R. Soc. London Ser. A.,* 294, 394, 1966.

106. **Kok, J. A., Poll, J. W., and Van Vroonhoven, C. G. E. M. M.,** Breakdown tests carried out on liquefied gases, *Appl. Sci. Res.,* B10, 257, 1962.

107. **Geballe, R. and Reeves, M. L.,** A condition on uniform breakdown in electron-attaching gases, *Phys. Rev.,* 92(4), 867, 1953.

108. **Bhalle, M. S. and Cragges, J. D.,** Measurement of ionization and attachment coefficients in sulfur hexafluoride in uniform fields, *Proc. Phys. Soc.,* 80, 151, 1962.

109. **Boyd, H. A. and Crichton, G. C.,** Uniform-field breakdown-voltage measurements in sulfur hexafluoride, *Proc. Inst. Electr. Eng.,* 119(2), 275, 1972.

110. **Chalmers, I. D. and Tedford, D. J.,** Spark breakdown in sulfur hexafluoride and arcton 12, *Proc. Inst. Electr. Eng.,* 118(12), 1893, 1971.

111. **Cookson, A. H., Farish, O., and Sommerman, G. M. L.,** Effect of conducting particles on AC corona and breakdown in compressed SF_6, *IEEE Trans. Power Appar. Syst.,* 91(4), 1329, 1972.

112. **Doepkin, H. C.,** Compressed-gas insulation in large coaxial systems, *IEEE Trans. Power Appar. Syst.,* 88(4), 364, 1969.

113. **Kouno, T.,** *Gaseous breakdown at Cryogenic Temperatures,* (in Japanese), Hoden-Kenkyu Discharge Study Group, No. 51, Tokyo, 12, 1973.

114. **Gerhold, J.,** Dielectric breakdown of helium at low temperatures, *Cryogenics,* October 1972, p. 370.

115. **Meats, R. J.,** Pressurized-helium breakdown at very low temperatures, *Proc. Inst. Electr. Eng.,* 119, 760, 1972.

116. **Careri, G.,** *Progress in Low Temperature Physics,* Vol. 3, North-Holland, Amsterdam, 1961, 58.

117. **Fowler, W. B. and Dexter, D. L.,** Electronic bubble states in liquid helium, *Phys. Rev.,* 176, 337, 1968.

118. **Atkins, K. R.,** Ions in liquid helium, *Phys. Rev.,* 116, 1339, 1959.

119. **Eggarter, T. P. and Cohen, M. H.,** Mobility of excess electrons in gaseous helium: a semiclassical approach, *Phys. Rev. Lett.,* 27, 129, 1971.

120. **Harrison, H. R. and Springett, D. E.,** Electron mobility variation in dense helium gas, *Phys. Rev. Lett.,* 35A, 73, 1971.

121. **Mathes, K. N.,** Dielectric properties of cryogenic liquids, *IEEE Trans. Electr. Insul.,* 2, 24, 1967.

122. **Swan, D. W. and Lewis, T. J.,** Influence of electrode surface conditions on the electrical strength of liquified gases, *J. Electrochem. Soc.,* 107, 180, 1960.

123. **Blaisse, B. S., van den Boogaart, A., and Erne, F.,** The electrical breakdown in liquid helium and liquid nitrogen, *Bull. Inst. du Froid, Paris,* 333, Annex 1958-1.

124. **Lehman, J. P.,** Measures dielectriques dan les fluides cryogeniques jusqu'á 200 kV-50 Hz, *Rev. Gen. Electr.,* 79, 15, 1970.

125. **Fallou, B., Bobo, J. C., Burnier, P., and Carvounas, E.,** Les Isolants Electriques Aux Tres Basses Temperatures, unspecified, 1971.

126. **Miura, T., Iwata, Y., and Furuto, Y.,** *Electrical Insulating Characteristics at Cryogenic Temperatures,* (in Japanese), Proc. 3rd Symp. EIM/IEEJ, Institute of Electrical Engineers of Japan, Tokyo, 1970, 57.

127. **Coelho, R. and Sibillot, P.,** Direct current pre-breakdown phenomena in liquid nitrogen, *Nature (London),* 221, 5182, 1969.

128. **Tanaka, T. and Okamoto, H.,** Discharge Phenomena in Cryogenic Electrical Insulation, Proc. 5th Int. Cryogenic Eng. Conf., Kyoto, 1974, 221.

129. **Tanaka, T.,** *Initiation of Internal Discharge in a Liquid-Nitrogen Filled Cavity,* Proc. 1976 IEEE Int. Symp. Electr. Insul., Montreal, IEEE, Piscataway, New Jersey, 1976, 128.

130. **Tanaka, T.,** Initiation of internal discharge in a liquid-nitrogen filled cavity, *IEEE Trans. Power Appar. Syst.,* 12(1), 35, 1977.

131. **Tanaka, T. and Okamoto, H.,** *Internal Discharges in a Liquid Nitrogen Filled Cavity (I),* (in Japanese), Tech. Rep. CRIEPI No. 73025, Central Research Inst. of Electric Power Industry, Tokyo, 1973, 1.

132. **Tanaka, T.,** *Internal Discharges in a Liquid Nitrogen Filled Cavity (II) — Pressure Effects,* Tech. Rep. CRIEPI No. 74056, Central Research Inst. of Electric Power Industry, Tokyo, 1975, 1.

133. **Tanaka, T.,** *Internal Discharges in a Liquid Nitrogen Filled Cavity (III) — Initiation,* Tech. Rep. CRIEPI No. 175015, Central Research Inst. of Electric Power Industry, Tokyo, 1975, 1.

134. **Zoleziowski, S. and Soar, S.,** Life curves of epoxy resin under impulses and the breakdown parameters, *IEEE Trans. Electr. Insul.,* 7(2), 84, 1972.

135. **Böttger, O.,** Heisse Elektronen in Isolierstoffen?, *Naturwissenschaften,* 59, 311, 1972.

136. **Bahder, G., Dakin, T. W., and Lawson, J. H.,** *Analysis of Treeing Type Breakdown,* CIGRE 15-05, CIGRE, Paris, 1974, 1.

137. **Tanaka, T.,** unpublished data.

138. **Tanaka, T. and Ikeda, Y.**, *V-t Characteristics of Polyethylene due to Internal Discharges*, (in Japanese), Tech. Rep. CRIEPI No. 73118, Central Research Inst. of Electric Power Industry, Tokyo, 1974, 1.

139. **Devins, J. C.**, *The Mechanism of the Formation of Discharges Limited by Series Dielectrics*, 1961 Ann. Rep. Conf. Elec. Insulation, National Academy of Sciences, Washington, D.C.

140. **Tanaka, T. and Ikeda, Y.**, *Discharge Magnitude and Statistical Time Lag in Internal Discharges*, (in Japanese), Tech. Rep. CRIEPI No. 73080, Central Research Inst. of Electric Power Industry, Tokyo, 1974, 1.

141. **Kalinichenko, I. S.**, Mathematical model for V-t characteristics of insulation, *Elektrichestvo*, No. 2, 55, 1973.

142. **Weibull, W.**, A statistical distribution function of wide applicability, *J. Appl. Mech.*, September 1951, p. 293.

143. **Tanaka, T. and Ikeda, Y.**, Internal discharge in polyethylene with an artificial cavity, *IEEE Trans. Power Appar. Syst.* 90(6), 2692, 1971.

144. **Tanaka, T. and Ikeda, Y.**, *Discharges in a Cylindrical Cavity (I) — Incompletely Sealed Cavities*, (in Japanese), Tech. Rep. CRIEPI No. 69057, Central Research Inst. of Electric Power Industry, Tokyo, 1970.

145. **Tanaka, T. and Ikeda, Y.**, *Discharge in a Cylindrical Cavity (II) — Sealed Cavities*, (in Japanese), Tech. Rep. CRIEPI No. 70037, Central Research Inst. Electric Power Industry, Tokyo, 1970, 1-25.

146. **Ikeda, Y. and Tanaka, T.**, *Effects of Gaseous Constituents on Internal Discharges in a Cylindrical Cavity (I)*, (in Japanese), Tech. Rep. CRIEPI No. 71018, Central Research Inst. of Electric Power Industry, Tokyo, 1971, 1.

147. **Ikeda, Y. and Tanaka, T.**, *Effects of Gaseous Constituents on Internal Discharges in a Cylindrical Cavity (II)*, (in Japanese), Tech. CRIEPI No. 72066, Central Research Inst. of Electric Power Industry, Tokyo, 1973, 1.

148. **Ikeda, Y. and Tanaka, T.**, *Effects on Gaseous Constituents on Internal Discharges in a Cylindrical Cavity (III)*, (in Japanese), Tech. Rep. CRIEPI No. 73008, Central Research Inst. of Electric Power Industry, Tokyo, 1973, 1.

149. **Ikeda, Y. and Tanaka, T.**, *Deterioration Test for Evaluation of Internal Discharges Resistance of Insulating Materials*, (in Japanese), Tech. Rep. CRIEPI No. 73085, Central Research Inst. of Electric Power Industry, Tokyo, 1974, 1.

150. **Heller, B. and Chlader, J.**, Der Einfluss der Ohm'schen und Dielektrischen Querfähigkeit auf den Koronavorgang im Festen Dielektrium, *Acta Tech. CSAV*, No. 1, 54, 1968.

151. **Chlader, J.**, Beitrag zur Problematik der Zeitdauer der Impulse bei den Koronaentladungen in Festen Dielektrium, *Acta Tech. CSAV*, No. 5, 621, 1971.

152. **Bailey, C. A.**, A study of internal discharges in cable insulation, *IEEE Trans. Electr. Insul.*, 2(3), 155, 1967.

153. **Salvage, B. and Tapupere, O.**, Measurement of current pulses due to discharges in a gaseous cavity in solid dielectrics, *Electron. Lett.*, 4(24), 529, 1968.

154. **Charters, J. S. T., Roaldset, S. A., and Salvage, B.**, Current pulses due to discharges in a gaseous cavity in solid dielectrics, *Electron. Lett.*, 6(18), 569, 1970.

155. **Mayoux, C. and Goldman, M.**, Partial discharges in solid dielectrics and corona discharge phenomena, *J. Appl. Phys.*, 44(9), 3940, 1973.

156. **Mayoux, C.**, *Corona Discharge and Aging Process of an Insulation*, Conf. Record of 1976 IEEE ISEI, IEEE, Piscataway, New Jersey, 1976, 276.

157. **Cribier, F., Goldman, A., and Lécuiller, M.**, *Analysis of the Species Created by Corona Discharges and Their Interactions with Surfaces*, Conf. Record of 1976 IEEE ISEI, IEEE, Piscataway, New Jersey, 1976, 282.

158. **Dakin, T. W. and Studniarz, S. A.**, *Voltage Endurance of Epoxy Resins with Microcavity Type Defects*, Conf. Record of 1976 IEEE ISEI, IEEE, Piscataway, New Jersey, 1976, 291.

159. **Kreuger, F. H.**, Determination of the internal discharge resistance of dielectric materials, *IEEE Trans. Electr. Insul.*, 3(4), 106, 1968.

160. **Kärkkäinen, S.**, *Internal Partial Discharges — Pulse Distributions, Physical Mechanisms and Effects on Insulations*, Tech. Res. Centre of Finland Publ. No. 14, 1976, 1.

161. IEEJ Committee on Corona Degradation of Insulating Materials, *Degradation of Insulating Materials by Corona Discharges (I)*, Institute of Electrical Engineers of Japan, Tech. Rep. No. I-74, Tokyo, 1966. 1.

161a. IEEJ Committee on Corona Degradation of Insulating Materials, *Degradation of Insulating Materials by Corona Discharges (II)*, Institute of Electrical Engineers of Japan, Tech. Rep. No. I-90, Tokyo, 1969, 1.

161b. IEEJ Committee on Corona Degradation of Insulating Materials, *Degradation of Insulating Materials by Corona Discharges (III)*, Institute of Electrical Engineers of Japan, Tech. Rep. No. I-106, Tokyo, 1973, 37.

161c. IEEJ Committee on Corona Degradation of Insulating Materials, *Degradation of Insulating Materials by Corona Discharges (IV)*, Institute of Electrical Engineers of Japan, Tech. Rep. No. II-43, Tokyo, 1976, 1.

162. **Toriyama, Y., Okamoto, H., Kanazashi, M., and Horii, K.,** Degradation of polyethylene by partial discharge, *IEEE Trans. Electr. Insul.,* 2, 83, 1967.

163. **Toriyama, Y., Okamoto, H., and Kanazashi, M.,** Breakdown of insulating materials by surface discharge, *IEEE Trans. Electr. Insul.,* 6(3), 124, 1971.

164. **Okamoto, H., Kanazashi, M., and Tanaka, T.,** *Deterioration of Insulating Materials by Internal Discharge,* IEEE PES Summer Meeting F 76 315 - 2, IEEE, Piscataway, New Jersey, 1976, 1.

165. **Anon.,** *Treeing in Organic Insulating Materials,* IEEEJ Committee Rep. I-No. 100, IEEE, Piscataway, New Jersey, 1971, 1.

166. **McMahon, E. J.,** A Tutorial on "Treeing", 147th Meeting of the Electrochemical Society, Toronto, Canada, 1975, 1.

167. **Olyphant, M., Jr.,** *A Compiled Study of Treeing in Insulation,* 3M Report, Minneapolis, 1963.

168. **Bahder, G., Dakin, T. W., and Lawson, J. H.,** *Analysis of Treeing Type Breakdown,* CIGRE Paper 15 - 05, CIGRE, Paris, 1974.

169. **Patsch, R.,** *Breakdown of Polymers: Tree Initiation and Growth,* Record of CEIDP Paper No. E6, National Academy of Sciences, Washington, D.C., 1975.

170. **Tanaka, T. and Greenwood, A.,** On Some Factors to Determine Tree Formation in Polyethylene, unpublished data.

171. **Tanaka, T.,** *Dielectric Breakdown of Polyethylene in Divergent Field,* Tech. Rep. of CRIEPI No. 176026, Central Research Inst. of Electric Power Industry , Tokyo, 1977, 1.

172. **Jocteur, R., Favrie, E., and Auclair, H.,** *Influence of Surface and Internal Defects on Polyethylene,* IEEE PES Summer Meeting F76 454-9, IEEE, Piscataway, New Jersey, 1976.

173. **Densley, R. J.,** Treeing in Cross-Linked Polyethylene Cables, Bull. REED, National Research Council Canada, Ontario, Vol. 22(1), 1, 1972.

174. **Robinson, D. M.,** The breakdown mechanism of impregnated paper cables, *Phys. Rev.,* 47, 90, 1935.

175. **Wartusch, J. and Wagner, H.,** *About the Significance of Peroxide Decomposition Products in XLPE Cable Insulations,* Conf. Rec. 1976 IEEE Int. Symp. EI, IEEE, Piscataway, New Jersey, 1976, 219.

176. **Noto, F. and Yoshimura, N.,** Proc. 11th Electr. Insul. Conf., Institute of Electrical Engineers of Japan, Tokyo, 1973, 137.

177. **Bahder, G. and Katz, C.,** *Treeing Effect in PE & XLPE Insulation,* Rec. Session III, CEIDP, National Academy of Sciences, Washington, D.C., 1972, 54.

178. **Izeki, N.,** Voltage Ageing and Acceleration Tests, *J. Inst. Electr. Eng. Jpn.,* 95(5), 20, 1975.

179. **Kreuger, F. H.,** *Câble Isolé au Polyéthylène À 'Imprégnation' Gazeuse,* CIGRE Paper 21-02, CIGRE, Paris, 1970.

180. **Kojima, T., Hanai, M., Yagi, K., Okusa, K., Aihara, M., and Haga, K.,** Characteristics of polyethylene impregnated with various gases, *IEEE Trans. Power Appar. Syst.,* 93(2), 579, 1974.

181. **Hayami, T.,** Development of liquid-filled type cross-linked polyethylene cable, *IEEE Trans. Power Appar. Syst.,* 88(6), 897, 1969.

182. **Kato, H., Maekawa, N., Inoue, S., and Fujita, H.,** *Effect and Mechanism of Some New Voltage Stabilizers for Crosslinked Polyethylene Insulation,* Ann. Rep. CEIDP, National Academy of Sciences, Washington, D.C., 1974.

183. **Fujiki, S., Furusawa, H., Kuhara, T., and Matsuba, H.,** *The Research in Discharge Suppression of High Voltage Crosslinked Polyethylene Insulated Power Cable,* IEEE PES Meeting 71 TP 195-PWR, IEEE, Piscataway, New Jersey, 1971, 1.

184. **Ashcraft, A. C., Eichhorn, R. M., and Shaw, R. G.,** *Laboratory Studies of Treeing in Solid Dielectrics and Voltage Stabilization of Polyethylene,* Conf. Rec. 1976 IEEE Int. Symp. Electr. Insul., IEEE, Piscataway, New Jersey, 1976, 213.

185. **Lever, R. C., MacKenzie, B. T., and Singh, N.,** Influence of inorganic fillers on the voltage endurance of solid dielectric power cables, *IEEE Trans. Power Appar. Syst.,* 92(4), 1169, 1973.

186. **Tanaka, T., Fukuda, T., Suzuki, S., Nitta, Y., Goto, H., and Kubota, K.,** Water trees in cross-linked polyethylene power cables, *IEEE Trans. Power Appar. Syst.,* 93(2), 693, 1974.

187. **Tanaka, T., Fukuda, T., and Suzuki, S.,** Water tree formation and lifetime estimation in 3.3 kV and 6.6 kV XLPE and PE power cables, *IEEE Trans. Power Appar. Syst.,* 95(6), 1892, 1976.

188. **Pohl, H. A.,** Some effects of non-uniform fields on dielectrics, *J. Appl. Physiol.,* 29, 1182, 1958.

189. **Matsuba, H. and Kawai, E.,** Water tree mechanism in electrical insulation, *IEEE Trans. Power Appar. Syst.,* 95(2), 660, 1976.

190. **Dakin, T. W.,** Electrical insulation deterioration treated as a chemical rate phenomena. I., *Trans. Am. Inst. Elec. Eng.,* 67, 113, 1948.

191. *Committee Report on Irradiation of Insulating Materials — Fundamentals,* (in Japanese), Institute of Electrical Engineers of Japan, Tokyo, Tech. Rep. I-74, 1966, 37.

192. *Committee Report on Irradiation of Insulating Materials — Application,* (in Japanese), Institute of Electrical Engineers of Japan, Tokyo, Tech. Rep. I-79, 1976, 1.

193. **Charlesby, A.,** *Atomic Radiation and Polymers,* Pergamon Press, Oxford, 1960.

194. *Guide for Type Test of Class IE Electric Cables, Field Splices, and Connections for Nuclear Power Generating Stations,* IEEE/ICC WG 12-32, and NPEC S/C 2.4, IEEE, Piscataway, New Jersey, P383, 1973.
195. **Sisman, O. and Bopp, C. D.,** ORNL-928, Oak Ridge National Laboratory, Oak Ridge, Tenn., 1951.
196. **Gumbel, E. J.,** *Statistics of Extremes,* Columbia University Press, New York, 1960.
197. **Weibull, W.,** A statistical distribution function of wide applicability, *J. Appl. Mech.,* September 1951, p. 293.
198. **Shiomi, H.,** *Introduction to Reliability Engineering,* (in Japanese), Maruzen, Tokyo, 1975.
199. **Kaneko, R. and Sugiyama, K.,** Proposals for testing method and value of statistical parameter, unpublished.
200. **Fischer, P. and Röhl, P.,** Application of statistical methods to the analysis of electrical breakdown in plastics, *Siemens Forsch. Entwicklungsber.,* 3(2), 125, 1974.
201. **Oudin, J. M., Rérolle, Y., and Thévenon, H.,** Théorie Statistique du claquage électrique, *Rev. Gen. Electr.,* 77(4), 430, 1968.
202. **Metra, P., Occhini, E., and Portinari, G.,** High voltage cables with extruded insulation — statistical controls and reliability evaluation, *IEEE Trans. Power Appar. Syst.,* 94(3), 967, 1975.
203. **Devaux, A., Oudin, J. M., Rérolle, Y., Jocteur, R., Noirclerc, A., and Osty, M.,** *Reliability and Development towards High Voltages of Synthetic Insulated Cables,* CIGRE Paper 21-10, CIGRE, Paris, 1968.
204. **Kaneko, R., Haruta, K., Sugiyama, K.,** Consideration on impulse breakdown stress in power cables by application of statistical method, (in Japanese), *Trans. Inst. Elec. Eng. Jpn.,* 93-B(4), 147, 1973.
205. **Matsuba, H.,** *Reliability Study of a System Consisting of a Cable and its Accessory,* (in Japanese), IEEJ-EIM-76-6, Institute of Electrical Engineers of Japan, Tokyo, 1976, 1.
206. **Tanaka, T. and Okamoto, T.,** Modified Weibull distribution — distributed lowest voltage model unpublished.
207. **Matsuba, H. and Maruyama, Y.,** *Deviation from the Weibull Distribution Function,* (in Japanese), Proc. 7th Symp. Electr. Insul. Mats., Institute of Electrical Engineers of Japan, Tokyo, 1974, 163.
208. **Lampert, M. A. and Mark, P.,** *Current Injection in Solids,* Academic Press, New York, 1970.
209. **Frank, R. I. and Simmons, J. G.,** Space-charge effects on emission-limited current flow in insulators, *J. Appl. Physiol.,* 38(2), 832, 1967.
210. **Tanaka, T.,** *Calculation of Ionic Mobility by the Voltage Reversal Method,* (in Japanese), Institute of Electrical Engineers of Japan, Tokyo, Rep. EI-44-9, 1969, 1.
211. **Tanaka, T.,** *Electric Field Distribution in Dielectrics Placed between Plane-Parallel Electrodes,* (in Japanese), Tech. Rep. CRIEPI No. 74004, Central Research Inst. of Electric Power Industry, Tokyo, 1974, 1.
212. **Tanaka, T.,** *Charge Injection into Solid Dielectrics,* (in Japanese), Tech. Rep. CRIEPI No. 74079, Central Research Inst. of Electric Power Industry, Tokyo, 1975, 1.
213. **Tanaka, T.,** *Charge Injection by Voltage Application into Polymer Dielectric Dielectrics,* IEEE PES Meeting Rec. A76 464-8, IEEE, Piscataway, New Jersey, 1976, 1.
214. **Oudin, J. M., Fallow, M., and Thévenon, H.,** Design and development of DC cables, *IEEE Trans. Power Appar. Syst.,* 86(3), 304, 1967.
215. **Occhini, E. and Maschio, G.,** Electrical characteristics of oil-impregnated paper as insulation for HVDC cables, *IEEE Trans. Power Appar. Syst.,* 86(3), 312, 1967.
216. **Tanaka, T.,** *Electrical Conduction in Oil-Impregnated Paper,* (in Japanese), Tech. Rep. CRIEPI No. 68094, Central Research Inst. of Electric Power Industry, Tokyo, 1969, 1.
217. **Counsell, J. A. H., Edwards, D. R., and Hartshorn, P. R.,** *Re-examination of the Variation with Stress of the DC Resistivity of Oil-Impregnated Paper,* Ann. Rep. CIEDP, National Academy of Sciences, Washington, D.C., 1966, 84.
218. **Tanaka, T.,** Effects of traps in polymeric materials on their electrical conduction and breakdown, (in Japanese), *J. Inst. Elec. Eng. Jpn.,* 86, 103, 1966.
219. **Tanaka, T.,** High field conduction in polyethylene, *Inst. Elec. Eng. Jpn.,* 89(4), 673, 1969.
220. **Tanaka, T. and Greenwood, A.,** Effects of charge injection and extraction on tree initiation in polyethylene, *IEEE Trans. Power Appar. Syst.,* 95(5), 1749, 1978.
221. **Shibuya, Y., Zoledziowski, S., and Calderwood, J. H.,** Void formation and electrical breakdown in epoxy resin, *IEEE Trans. Power Appar. Syst.,* 96, 198, 1979.

Chapter 3

TESTING

3.1. TESTING OF TAPED CABLES

An adequate test program is necessary for power cables before installation and during their operation. Testing can be conveniently classified into three categories:[1-3]

- Factory tests
- Field tests (installation tests)
- Maintenance tests

A variety of test procedures have been proposed, some of which are being followed. The precise details of which tests to select and how to conduct the tests vary from country to country. The first two categories itemized above are generally made by the manufacturer. Details of U.S. test procedures are specified in AEIC No. 4-69,[4] for impregnated-paper-insulated low-pressure oil-filled cables, AEIC No. 2-67,[5] for impregnated-paper-insulated high-pressure pipe-type cables, and AEIC No. 5-75,[6] for polyethylene and cross-linked polyethylene insulated shielded power cables rated 5 through 69 kV.

3.1.1. Factory Tests

Factory tests are carried out prior to shipment to assure that the cables as manufactured meet all the conditions required by the purchaser. They often consist of type tests, acceptance tests, and special tests. The performances of a specific cable design and its quality level are verified by type tests. It is therefore unnecessary to repeat the tests unless any specification is changed. Tests include cable components as well as the finished cable itself. Manufacturers conduct acceptance tests or routine tests on all cables to be shipped in order to confirm the soundness of their performance and to give a quality guarantee. Special items may be investigated for a finished cable if both the manufacturer and the purchaser agree.

A. Factory Frame Tests on OF and POF Cables

AEIC specifications No. 4-69 and No. 2-67 set out the tests which should be carried out before shipment. Any cable which fails to meet the requirements of the following tests cannot be shipped without the specific approval of the purchaser.

Conductor resistance — The conductor resistance of each factory or shipping length of cable shall be measured by a wheatstone bridge or other similar means. The temperature of the cable at the time of the measurement may be determined by a thermometer, the bulb of which must be in contact with the lead (OF cable) or the sheath or outer covering (POF cable), and be suitably protected from the air, or by other similar means. If the temperature differs from 25°C, the measured resistance shall be corrected to 25°C by multiplying by the appropriate factor taken from Table 3.1.1. The Japanese standard for temperature is 20°C.

High voltage test — Each factory or shipping length of lead-covered OF cable shall be subjected to a high voltage test at an AC voltage corresponding to 300 V/mil (12 kV/mm) of the specified insulation thickness; it should be maintained for 15 consecutive minutes. Less than half voltage shall be impressed on POF cable, at atmospheric pressure, for 15 consecutive minutes, as listed in Table 3.1.2. for typical nominal voltages. The initially applied voltage shall not be greater than 20% of the test voltage. The rate of increase shall be approximately uniform and such that the specified test voltage is attained in not less than 10 nor more than 60 sec.

Table 3.1.1.
TEMPERATURE
CORRECTION FACTORS
FOR CONDUCTOR
RESISTANCE

Temperature of copper at time of measurement (°C)	Factor[a]
0	1.107
5	1.084
10	1.061
15	1.040
20	1.020
25	1.000
30	0.981
35	0.963
40	0.945
45	0.928
50	0.912

Note: Table is based on the resistance-temperature coefficient of copper of 100% conductivity (International Annealed Copper Standard) viz., 0.00385 at 25°C.

[a] Factors for temperatures between those shown shall be obtained by linear interpolation.

Table 3.1.2.[4,5]
HIGH VOLTAGE FACTORY TESTS TO BE MADE ON EACH LENGTH OF CABLE

Rated voltage (kV) phase-to-phase	Size of conductors AWG or kcmil	Insulation thickness (mil)	OF cable test voltage (kV)	POF cable[a] test voltage (kV)	Ratio POF/ OF
69	4/0 to 2500 (OF) 3/0 to 3000 (POF)	250	75.0	42.7	0.49
138	250 to 2500 (OF) 3/0 to 3000 (POF)	505	151.5	75.7	0.50
230	600 to 2500 (OF) 500 to 3000 (POF)	760	228.0	95.0	0.42
345	1000 to 2500 (OF) 1000 to 3000 (POF)	1035	310.5	104.0	0.33

Note: 100 mil = 2.54 mm; 2000 kcmil = 1000 mm².

[a] To be tested at atmospheric pressure.

Ionization test — The ionization factor is defined as the change in power factor of the dielectric with change of electrical stress. It is the difference in power factor of the dielectric at power frequency, measured at an average stress of 100 V/mil, and the power factor measured at the stress at which the power factor is a minimum, as shown by a power factor-voltage curve which is typical of the cable under test. A Schering bridge is normally used to measure the power factor. Each factory or shipping length of OF cables shall be tested to determine the ionization factor. For POF and GOF (gas-pressurized OF) cables (medium viscosity), the complete length shall be tested at atmospheric pressure. If the ionization factor, corrected to 25°C, exceeds 0.1% or the value stated by the manufacturer in his proposal, the length of cable shall be considered to have failed to meet the requirements of the specification. High-viscosity, gas-pressure cables shall be subjected to the test at any pressure from atmospheric to 12 psig (0.84 kg/cm^2 G); the specified high and low stress shall be 80 V/mil (3.2 kV/mm) and 40 V/mil (1.6 kV/mm), respectively, and the ionization factor shall not exceed 2.0.

Spark test for OF cables — The manufacturer shall-spark test the entire covering of every length of cable having a corrosion protective covering during the manufacturing process. The covering shall withstand the voltage indicated in Table 3.1.3. The voltage shall be applied between an electrode at the outside surface of the covering and the sheath, for not less than 1 sec.

Thickness of Sheath — The average and the minimum thickness of the lead sheath of an OF cable and the extruded sheath other than lead for POF cable shall be determined from a specimen removed from each end of each shipping length of cable. The thickness of the sheath shall be determined by measurements made with a micrometer caliper directly on specimens of sheath removed from the cable. At least five separate measurements, approximately equally spaced around the circumference, shall be made on each specimen. When the lead sheath of OF cable is found to be deficient in thickness, the sheath shall not be repaired, but the lead may be stripped from the entire length of the cable before impregnation and the cable released. In case of an extruded sheath of a POF cable, the average thickness shall not be less than the specified thickness and the minimum thickness shall not be less than 90% of the specified thickness.

Oil-check test for OF cable — At least 5 hr but preferably 15 hr, if production conditions permit, after the cable has been completed and prepared for shipment, two samples of oil shall be taken from each length of cable, one through the pulling eye and one from the reel reservoir. The power factor of the samples shall be determined in accordance with the American Society for Testing and Material (ASTM) D 924. The power factor shall be measured at power frequency and at 100°C. The dielectric strength of the samples shall be determined in accordance with ASTM D 877. If the power factor is greater than 1.0%, or the dielectric strength is less than 28 kV, the length of cables from which the sample was withdrawn shall be considered to have failed the test.

Expulsion test for OF cable — After each length of cable has been sealed and prepared for shipment, it shall be tested for entrapped gas by a special technique described in AEIC No. 4-69. The expulsion constants calculated from the test results shall not exceed 0.065.

B. Factory Sample Tests on OF and POF Cables

The following AEIC specifications describe tests on samples.

Mechanical integrity — The mechanical integrity test shall be made by the manufacturer on samples selected by him with such frequency that there will be 1 examination for each 5000 ft (1500 m) of cable or major fraction thereof, on the order, but at least 1 examination for each order. Test items are specified as follows:

1. Visual inspection of at least 1-ft long sample of cable carefully cut in order to check any mechanical injury
2. Determination of the average and minimum insulation thickness. The average thickness of the insulation at any part of the cable shall not be less than the specified thickness. The minimum thickness at any part of the cable shall not be less than 90% of the specified thickness, but the difference between specified and minimum shall not exceed 50 mil (0.127 mm)

Table 3.1.3.
AVERAGE THICKNESSES OF PROTECTIVE COVERINGS AND TEST VOLTAGES

Calculated Approximate Diameter over Sheath (inches)	Thickness (mils)		Spark Test Voltage (kV, ac)
Polyethylene Covering			
0 to 1,500	80		6.5
1,501 to 2,250	90		8.0
2,251 to 3,000	110		10.0
Over 3,000	125		11.5
Neoprene Covering	Neoprene compound	Total covering	
0 to 1,500	40	80	3.0
1,501 to 3,000	60	110	4.0
Over 3,000	60	140	5.0

Note: 1.000 in. = 25.4 mm., 100 mil. = 2.54 mm.

3.   Examination of the condition of the binder and the outer shielding tapes
4.   Examination of the number of registrations and the presence of wrinkles and tears in the insulation. Not more than two adjacent tapes in any consecutive layers shall be registered per foot of cable. The total number of registrations in the conductor insulation shall not exceed 1 per 15 tapes or fraction thereof. If a sample shows a degree of wrinkling or softness which may impair the dielectric strength of the cable, a length with approximately the same degree of wrinkling or softness shall be subjected to a high-voltage withstand test as described previously
5.   Bending ability of the sheath of OF cable which can be determined with an IPCEA cable sheath bending machine. Complete fracture of the specimen at any point around the circumference shall not occur in less than 22 90° bends

Bending test — The sample selected for the bending test shall not be less than 10-ft (3 m) long nor more than 20-ft (6 m) long. The diameter of the form around which the sample is to be bent shall be equal to that shown in Table 3.1.4. except that in no case shall the diameter for single-conductor cable be less than 20 times the bare conductor diameter. The overall diameter of the POF cable includes skid wires, but where cable is not round, the smallest diameter shall be used. The cable shall be bent 180° around a cylindrical form of the stated diameter and then bent 180° in the opposite direction around the same or a similar form and again straightened. This cycle of operation shall be performed twice, making a total of four bending operations. The cable shall be so held during the operations that it cannot revolve around its own axis. In case of OF cable, prior to this test, it shall be drained of oil and then subjected to a temperature below 10°C for not less than 2 hr. The ends of the sample shall not be sealed. The following items shall be examined:

1.   Crack formation: the sheath shall show no cracks after the cable is straightened prior to the last bending operation.

Table 3.1.4.
BENDING RADIUS OF CABLE FOR TESTING

OF Cable		Ratio O.D. of cylindrical form to O.D. of cable	POF cable
Number of conductors	Insulation (mil)		
1	500 or less	12	350 mil or less
1	Over 500	16	351 to 800 mil
3	Any	12	—
—	—	20	More than 800 mil

Note: 100 mil = 2.54 mm.

2. Deformation: after the sample has been bent for the 4th time and while still bent, 3 adjacent pieces of cable, each approximately 1-ft (30 cm) long, shall be cut from the center of that portion of the sample which has been subjected to the most bending. On the center 1-ft piece removed, a deformation measurement shall be made before it is dissected. The minimum thickness of the insulation shall be measured at each end of the specimen. The sample shall be considered to have failed the test if the minimum thickness of the insulation is less than 80% of either the specified thickness or the measured average thickness of an unbent specimen cut adjacent to the bent specimen examined.

3. Torn paper: the overall covering or the lead sheath shall be stripped from all three pieces and the insulation removed, not more than five tapes at a time. There shall be no more than two adjacent tapes torn at the same spot. The total number of tapes torn in any 1 ft of conductor insulation of the 3 ft of cable examined shall not exceed 2 in any 10 consecutive tapes.

4. Thickness of the lead sheath of OF cable: any thickness out of 10 point measurements shall be not less than 80% of the specified thickness.

Tests on the corrosion protective covering of OF cable — The manufacturer shall make tests on samples for the applicable type covering. The tests are to be conducted on representative samples not less than 1 ft in length. Items to be examined are

1. Thickness
2. Tightness of application of the polyethylene covering to the sheath
3. Mechanical tests on neoprene covering

High-voltage time test — The sample selected for this test shall be not less than 30-ft (9 m) long. This sample may, at the option of the manufacturer, be used for the dielectric power loss and power factor tests before the high-voltage time test is made. The high-voltage time test shall be performed at room temperature with OF cable under a positive oil pressure, with POF or medium viscosity GOF cable installed in a pipe or other container under a pressure not exceeding 100 psig (7 kg/cm² G) with specified pressure medium, and with high-viscosity GOF cable, under a gas pressure not exceeding 150 psig (10.6 kg/cm² G). A test voltage corresponding to 430 V/mil (17 kV/mm) of insulation thickness shall be applied for a period of 6 hr.

Dielectric power loss and power factor tests — The sample selected for this test shall be not less than 10 ft (3 m) long. The dielectric loss per foot and the power factor of the dielectric shall be determined at the rated voltage of the cable and at not less than four temperatures. One temperature shall be room temperature and the others shall be 80°C, 90°C, and 105°C. The measurements at 105°C are for engineering information only. There are no specification limits for power factor at this temperature. The measured power factor shall not exceed the values as shown in Table 3.1.5.

Table 3.1.5.
MAXIMUM PERMISSIBLE POWER
FACTOR

Temperature of cable[a] (°C)	Maximum power factor (%)	
	15 to 161 kV	Over 161 kV
Room to 60	0.50	0.30
80	0.55	0.25
90	0.60	0.28
105[b]	—	—

[a] Uniform temperature throughout cable.
[b] Measurements for engineering information only.

Ionization factor test for POF and GOF cables — A test for ionization factor shall be made as indicated in a frame test except that it shall be at a pressure not exceeding 200 psig (14 kg/cm² G) in the specified pressure medium. The stresses and ionization factors shall be as indicated in Table 3.1.6.

3.1.2. Installation Tests

It is necessary to conduct a completion test or an installation test prior to service in order to confirm whether or not the cable system which has been completed meets the requirements specified in appropriate regulations and/or specifications and agreed upon by the manufacturer and the purchaser. In Japan some of the test items are conducted under the supervision of the Ministry of Trade and Industry. The purchaser may make such of the following tests in the U.S. as he may elect.

A. High Voltage Withstand Test

1. Proof tests at a voltage not exceeding 225 V/mil (9 V/mm) of insulation thickness, applied for 5 consecutive minutes during both installation and service periods
2. Installation-acceptance test either at 190 V/mil (7.6 kV/mm) of insulation thickness applied for 4 consecutive hours, or at 255 V/mil (9 kV/mm) of insulation thickness for 15 consecutive minutes after installation and before the cable is placed in regular service
3. Any of the above high-voltage tests may be made with direct potential instead of alternating potential, in which case the test voltage for the cable shall be 2.4 times the prescribed alternating potential, provided the temperature of the cable does not exceed 25°C (77°F). If the temperature exceeds 25°C, the factor, 2.4, shall be reduced by 0.013 for each degree in excess of 25°C

3.1.3. Some Considerations on Impulse Tests

High voltage tests are made on the basis of the specification recommended by the International Electrotechnical Commission (IEC) in Europe, on the basis of AEIC specification in the U.S., and on the basis of Japanese Cable Standards and individual purchaser's specification in Japan.[7] The IEC specification prescribes testing methods only, while the others specify cable structure, size and materials, and testing methods.

Power cable design entails long-term stability against lightning surges, sustained over voltage and service voltage. Judging from the withstand voltage characteristics of OF cable, its impulse voltage withstand performance is certainly critical to its design. Each of the stated specifications may therefore be characterized by the systems approach toward the level of impulse voltage to be allowed in the cable system and a suitable safety factor to be set for the cables.

Table 3.1.6.
MAXIMUM IONIZATION FACTOR

Rated voltage (kV)	Average stress, volts per mil of specified insulation thickness		Maximum ionization factor (%)
	High stress	Low stress	
15 to 68	175	20	0.1
69 to 115	185	20	0.1
116 to 161	190	20	0.1
Over 161	230	20	0.05

Note: 100 V/mil = 4 kV/mm.

Most present impulse voltage test specifications describe methods only, they lack a description of their objective or purpose. In any event, this must be to assure the integrity of the cable against intrusion of lightning surges and the incidence of switching surges. The specified impulse test voltage level is often determined in reference to the basic impulse level (BIL) for transformers. It appears to be defined as 75% or 100% of the BIL in Europe, these being the values for cable system proper. In Japan it is 100 to 120% (130 to 150% in part). In the U.S. it is 100% of the BILs of transformers or such values as coordinate with the flashover voltage of overhead transmission lines.

As mentioned already, the AEIC specification does not include any impulse voltage tests, because insulation thickness is specified for each rated voltage. The insulation thickness is determined so as to withstand at least the BIL, i.e., an average stress of 1300 V/mil (52 kV/mm). The remaining specifications (other than AEIC) require that the standard impulse voltage be applied to cables after they are bent. The impulse voltage test may be made either at room temperature or at elevated temperatures, as called for by individual specifications. The number of impulse applications is 10 to 20 in Europe and 3 in Japan.

3.1.4. Development Tests for 500 kV or Higher Voltage Taped Cables

The recent need to develop 500 kV and higher voltage cables has led to the establishment of central stations in which utilities, manufacturers, and even people from different countries may collaborate for an agreed purpose. This has come about for both technical and economic reasons. The Waltz Mill cable testing station in the U.S., the Takeyama cable testing site in Japan, the CESI UHV power cable testing station in Italy, and the Renardieres underground cable testing site in France are all examples of this approach.

A. The Waltz Mill Cable Testing Station[8,9]

The Waltz Mill test facility is part of the Electric Power Research Institute (formerly the Electric Research Council) program for research on underground transmission. This facility, located on a 60-acre (240,000 m²) site, is designed to perform accelerated life tests and other evaluations on single-phase samples, ranging in operating voltage from 115 kV to 800 kV. The station is divided into two test areas, the HV area and the EHV area. Up to six samples can be installed in each area. In the HV area, samples rated 115, 138, and 230 kV can be tested, while the EHV area is suitable for testing samples rated 345, 500, and 750 kV. The voltage source for the EHV area comprises 3 single-phase 92/183 kV regulators and 3 single-phase 183/1150 kV autotransformers, 1 rated 40 MVA and the remaining 2 rated 20 MVA. Suitable taps and load tap changing mechanisms provide test bus voltages from 300 to 1100 kV (line-to-line) with individual separate phase regulation. The voltage source for the HV area consists of 3 single-phase regulators and 3 single-phase 5 MVA autotransformers which provide test bus voltages from 115 to 345 kV.

Both continuous and cyclic operating modes are employed. During continuous loading periods, conductor temperatures are maintained at the desired value by a combination of circulating current and pipe heating. Circulating currents are induced in the conductor by current transformers at the cable risers. Cyclic loading first heats the cable for 12 hr to attain the desired conductor temperature, then allows 12 hr for cooling. During the 12-hr cooling period, both pipe heating and ciruclating currents are discontinued, but the applied test voltage remains. Pipe heating and load current are computer controlled.

Dielectric dissipation factor is the primary quantity monitored to verify the performance of paper-insulated cables. A transformer ratio-arm bridge and a modified Schering bridge system are equipped for this purpose. Partial discharge measurements are made with a differential corona bridge.

A computer is used as an integral part of the instrumentation system. In addition to the control and monitoring function the computer also collects and processes data on pipe, shield, conductor, pothead, and ambient temperatures; on bus voltage and circulating current; on pipe pressures, oil flow rate, and on wind velocity.

Since 1969, the facility has been used for 500 kV and higher voltage POF cable evaluation, which followed the Cornell project on 345 kV OF and POF cables completed in 1966, and for 138 kV and 230 kV XLPE cable development. The capabilities of this station are not limited to this work. Prototypes of resistive cryogenic, superconducting, gas insulated, and other underground transmission systems with rated voltages from 115 kV to 750 kV can be tested.

B. CESI UHV Cable Testing Station

CESI is an independent organization founded in 1956 at the initiative of a group of Italian electrical and electromechanical manufacturers and all the Italian utility companies. It has since been absorbed by ENNEL, the Italian Electrical Power Board. CESI is located in Milan, Italy. A new plant was completed in 1974 for evaluation test programs on UHV power cables.[10]

The testing plant has 2 main parts: 1 is the external area (a square platform 25 × 25 m) on which the power circuit is installed together with the cable sealing ends and the insulated support for the compensating capacitors, and the other is the tunnel (8 × 77 × 2.5 m free height) in which cables, joints, and minor components of the circuit can be housed with the principal purpose of reducing atmospheric effects.

Voltage is provided by an HV (800 kV 20 MVA) transformer with two step-by-step adjustable reactors. All parts of the circuit on the high voltage side are suitably shielded by means of ''polycon type'' electrodes which make them corona-free up to a maximum voltage of 800 kV rms. A system of guard wires protects the entire external area against lightning. Cable heating can be provided by current flow through the cable conductors and be suitably placed external resistance heaters which are automatically controlled. Temperatures on cable conductors sheaths, and joints are measured. A bank of capacitors is employed to compensate for the inductance of the cable loop under test.

Dissipation factors of accessories and cable sections, relevant oil pressures and temperatures, together with the voltage, are the major quantities to be measured throughout testing period. Oil pressures are measured directly by means of pressure gauges. A remote-controlled pneumatic connector system is employed.

3.2. TESTING OF EXTRUDED CABLES

Apart from maintenance tests, the tests specified for extruded cables in AEIC No. 5-75[6] consist of factory tests and installation tests as in the case of taped cables. The factory tests include qualification tests (or type tests), production sampling tests, and tests on completed cable as well as material tests which are peculiar to this type of cable. The installation tests are made during and after installation.

Table 3.2.1.
MAXIMUM ALLOWABLE SIZES OF IMPERFECTIONS IN EXTRUDED CABLE INSULATION

	AEIC specification		Japan specification	
Rated voltage	(5 ~ 69 kV)		11 ~ 13 kV	66 ~ 77kV
Insulation	PE	XLPE	XLPE	XLPE
Projection from semiconducting layers	250 μm	250 μm	250 μm	250 μm[a]
Voids	50 μm	250 μm	80 μm	80 μm
Contaminants	250 μm	250 μm	250 μm	250 μm[a]
Translucents	1.25 mm	1.25 mm	—	—

[a] It is considered that it should be reduced into 100 μm in the future.

3.2.1. Factory Tests
A. Insulation, Voids, and Contaminations

As described in Volume II Sections 1.1.4. and 1.1.5., imperfections in extruded insulation are critical to its dielectric performance. They are therefore categorized for polyethylene and cross-linked polyethylene in the following manner.

High-molecular weight polyethylene compound is used as insulation in "polyethylene-insulated" cable. The insulation of the completed cable shall be free from:

1. Any void larger than 2 mil (50 μm)
2. Any contaminant larger than 10 mil (250 μm) in its largest dimension. The number of contaminants of sizes between 2 and 10 mil (50 and 250 μm) shall not exceed 15/in.[3] of insulation (0.92/cm[3]) for all voltages
3. Any translucent material that is larger than 50 mil (1.25 mm) in its radial vector projection

Cross-linked polyethylene insulation of a completed cable shall be free from:

1. Any void larger than 5 mil (125 μm) (for cables rated 5 through 69 kV). The number of voids larger than 2 mil (50 μm) shall not exceed 30/in.[3] of insulation (1.84/cm[3]) for all voltages
2. Any contaminants (opaque material or material that is not homogeneous cross-linked polyethylene) larger than 10 mil (250 μm) in its largest dimension. The number of contaminants of sizes between 2 and 10 mil (50 and 250 μm) shall not exceed 15/in.[3] of insulation (0.92/cm[3]) for all voltages
3. Any translucent material that is larger than 50 mil (1.25 mm) in its radial vector projection

"Filled cross-linked polyethylene" contains 10% or more of mineral fillers by weight. An unfilled compound may contain up to 21/2% carbon black for use at 5 kV and no carbon black above 5 kV.

The contact area between the insulation and conductor shield shall have no projections or irregularities which extend more than 10 mil (250 μm) from the cylindrical surface of the conductor shield.

For comparison, Table 3.2.1. lists maximum allowable sizes of imperfections in extruded cable insulation as specified by AEIC and IEEJ.[11] Translucents are considered to be amber or oxidized polyethylene. They are not included as contaminants in the IEEJ specification.

Table 3.2.2.
BIL VALUES
CORRESPONDING TO
CABLE RATING

Cable rating (kV)	BIL (kV)
5	60
8	95
15	110
25	150
28	150
35	200
46	250
69	350

B. Qualification Tests

Qualification tests or type tests are intended to demonstrate the manufacturer's capability to furnish high quality cable with the desired performance characteristics. Such tests comprise a high-voltage time test, an impulse breakdown test, a cyclic aging test, and a resistance stability test.

High-voltage time test — A test voltage of 200 V/mil (8 kV/mm) shall be applied and held for a period of 4 hr. The voltage shall be increased and held for 1 hr. each at voltages of 300 and 360 V/mil (12 and 14.4 kV/mm). The voltage is then to be increased in 40 V/mil (1.6 kV/mm) steps and held 1/2 hr at each value, continuing to cable breakdown. The test may be discontinued for 15 kV rated cable only by reason of a cable failure outside the terminations, and for cable rates 25 through 69 kV by reason of a termination failure only, if the 480 V/mil (19.2 kV/mm) step has been concluded. The cable under test must withstand the 4-hr test at 200 V/mil (8kV/mm) and the 1st hour at 300 kV/mil (12 kV/mm).

Impulse breakdown test — An impulse breakdown test shall be performed with the temperature of the conductor at a nominal 25°C and 130°C in case of cross-linked polyethylene. The 90°C and 130°C shall be achieved by circulating current in the conductor. Three impulses of both polarities with a magnitude equal to the BIL shown in Table 3.2.2.,[6] shall be applied. The voltage is then to be raised in steps of approximately 25% of BIL with 3 impulses of negative polar polarity applied at each step, continuing to cable breakdown outside the terminals. A cable shall withstand an impulse voltage of 125% of the listed BIL rating at room temperature.

Cyclic aging test — A 42-day aging test, consisting of 2 periods, shall be carried out. During the 1st 21-day period, an AC voltage of twice rated voltage to ground is continuously applied. Contemporaneously, such load current as will produce a nominal conductor temperature of 75°C for PE and 95°C for XLPE is passed through the conductor for approximately 8 continuous hours, but not less than 6 hr, each working day. During the 2nd 21-day period, a voltage of 3 times rated voltage is ground is continuously applied and such load current as will produce a conductor temperature of 90°C for PE and 130° C for XLPE is circulated under conditions similar to those just mentioned. Partial discharge level, insulation thickness, and power factor shall be measured prior to the test and at the end of each 7-day test period. After the test is completed, the cable shall be subjected to a physical examination which includes measurement of the volume resistivity of the conductor and insulation shields and the insulation tensile strength and elongation at rupture.

Resistance stability test — The resistance of the extruded insulation semiconducting covering on a 1-ft sample shall be measured after 24 hr at 90°C for the thermoplastic covering, or at 110°C for the thermosetting covering, and at 14-day intervals from the start of the test until resistance stability is reached or its value exceeds 50,000 Ω-cm.

Table 3.2.3.
INSULATION THICKNESS, TEST VOLTAGES, AND CONDUCTOR SIZES
FOR POLYETHYLENE AND CROSS-LINKED POLYETHYLENE
INSULATED POWER CABLES

Rated voltage phase-to-phase (kV)	Conductor size (AWG or kcmil)	Minimum average insulation thickness (mil)		AC test voltages (kV)		DC test voltages (kV)	
		A	B	A	B	A	B
5	8 to 1000	90	115	13	16	35	45
	Above 1000	140	140	13	16	35	45
8	6 to 1000	115	140	18	22	45	55
	Above 1000	175	175	18	22	45	55
15	2 to 1000	175	220	27	33	70	80
	Above 1000	220	220	27	33	70	80
25	1 to 2000	260	320	38	48	100	120
28	1 to 2000	280	345	42	49	105	125
35	1/0 to 2000	345	420	49	63	125	155
46	4/0 to 2000	445	580	66	86	165	215
69	500 to 2000	650	650	98	98	245	245

Note: There are two insulation thicknesses specified in some cases for polyethylene and cross-linked polyethylene insulated cable having the same voltage rating. Columns A and B provide a choice of these thicknesses for the 100% insulation level. This choice will depend on the pertinent operating experience and evaluation of other factors such as conductor size, initial cost, replacement cost, impulse strength, etc. (After additional operating and testing experience is accumulated, it is anticipated that Columns A and B can be replaced by a single column that will be generally acceptable to the utility industry.) When the 133% insulation level is required, Column B, insulation thickness may be specified.

C. Production Sampling Test

In addition to the stated imperfection test, a heat distortion test, a solvent extraction test, a structural stability test, and a dimensional stability test shall be made.

D. Tests on Completed Cable

Electrical performance tests on completed cable consist of AC and DC voltage withstand tests and a partial discharge measurement. Each insulated conductor in the completed cable shall be tested at the voltages given in Table 3.2.3.,[6] the details of which are specified in IPCEA S-61-402 for PE and in IPCEA S-66-524 for XLPE. Multiplexed cable shall be tested after multiplexing

Each reel of completed shielded power cable shall comply with the maximum partial discharge in picocoulombs specified in Table 3.2.4.[6]

3.2.2. Installation Tests

A. Test Voltages and Their Application Time

DC overvoltage tests are specified instead of AC overvoltage tests for field test. DC test voltages are 2.5 to 2.8 times the AC test voltage values. AEIC No. 5-75 specifies tests during and after installation of extruded cables as follows:

1.   At any time during installation, a DC proof test may be made at a voltage not exceeding 75% of the DC test voltage specified in Table 3.2.3. applied for 5 consecutive minutes.
2.   After installation and before the cable is placed in regular service, a high voltage DC test may be made at 80% of the specified DC test voltage applied for 15 consecutive minutes.

Table 3.2.4.
MAXIMUM PERMISSIBLE PARTIAL DISCHARGES
FOR EXTRUDED CABLE

Ratio (V_T/V_{RG})					
Maximum permissible partial discharge in picocoulombs — all voltages		1.5	2.0	2.5	3.05
		5	20	35	50
Cable voltage rating (kV)	Line to ground (V_{RG})	Test voltages[a] (V_T) in kV corresponding to (V_T/V_{RG}) ratio			
5	2.9	4.3	5.8	7.2	8.6
8	4.6	6.9	9.2	11.5	13.8
15	8.7	13.0	17.3	21.6	26.0
25	14.4	21.6	28.8	36.0	43.2[a]
28	16.2	24.2	32.3	40.4	48.4[a]
35	20.2	30.3	40.4	50.5[a]	60.6[a]
46	26.5	39.8	53.1	66.3[a]	— [a]
69	40	60	80	— [a]	—

Note: The maximum partial discharge in picocoulombs shall not exceed the values given by the equation

$$pC = 5 + \left(\frac{V_T}{V_{RG}} - 1.5 \right) 30$$

where V_T = test voltage, and V_{RG} = rated voltage, phase to ground. The formula applies when the quantity in parentheses is not less than zero.

[a] The AC test voltages in Table 3.2.3. for the specified walls shall not be exceeded.

3.   After the cable has been completely installed and placed in service, a DC proof test may be made at any time during the period of the guarantee at 65% of the specified DC voltage applied for 5 consecutive minutes.

B. New DC Voltage Tests in the Field

A new DC test level formula, prescribing 70% of rated system BIL rounded off to the nearest 5 kV, has been proposed by the IEEE Insulated Conductor Committees (ICC). This proposal also includes the DC voltage test during service. The test voltages are shown in Table 3.2.5.[12] Class I and Class II correspond to the normal reliable cable system and to the highly reliable cable system, respectively. The 70% of BIL initial test voltage would be continued for in-service tests of Class II cables, but for Class I cables, in-service testing would be conducted at 75% of the initial test voltage.

3.3. MAINTENANCE TESTS

Periodic maintenance tests are made with the object of preventing failure of cable systems during operation. Several methods for investigating insulation are listed and briefly interpreted in Table 3.3.1. They can be roughly classified into the DC high voltage method, the partial discharge measurement method, and the loss tangent measurement method. One should not expect a single method to properly detect defects in a cable; a suitable comprehensive approach which includes the measurements of various representative parameters is necessary.

Table 3.2.5.
PROPOSED IEEE MAXIMUM DC
TEST POTENTIALS FOR ALL CABLE
TYPES

System voltage (kV)	BIL (kV)	Initial install. (kV)	Service Class I (kV)	Class II (kV)
2.5	60	40	30	40
5	75	50	40	50
8.7	95	65	50	65
15	110	75	55	75
23	150	105	80	105
28	170	120	90	120
34.5	200	140	105	140
46	250	170	130	170
67	350	245	185	245

Note: 15 min duration — 100% insulation level (I.L.).

From Lee, R. H., *New Developments in Cable System Testing*, IEEE Conf. Rec. 10th Annual Meet. Ind. Appl. Soc., IEEE Piscataway, New Jersey, 1975, 822. With permission.

3.3.1. Electrical Tests

A. DC High Voltage Method

This method judges the cable insulation performance from the temporal change in current when a DC voltage is applied between cable conductor and sheath (Table 3.3.1.). The current generally comprises three components: the displacement current which is large but decreases dramatically a short time after the DC voltage is applied, the absorption current which decreases gradually over a comparatively long time, and the leakage current which does not change with time.

The absorption current in the second step of current response is accepted as an indicator of a certain insulation performance in power cables; it is expressed by several measures such as the polarity index, the interphase imbalance coefficient, the weak-spot ratio, and the current kick. The polarity index is the ratio of the current 1 min after voltage application to the current at a specified later time (7 to 10 min after voltage application). There is agreement that the polarity index tends to decrease and the leakage current increase as oil-impregnated paper insulation deteriorates. This measure, however, seems to be ineffective for extruded cable insulation.

The interphase current imbalance is given by the ratio of the maximum to minimum difference in the leakage currents of three phases to their average leakage current. This appears to be an effective way of determining which of the three phases has deteriorated most.

In oil-impregnated paper (solid) cable and butyl-insulated cable, it is recognized that the characteristic of insulation resistance vs. applied voltage has an inverted U shape when the insulation is sound. But as it deteriorates, the peak of insulation resistance tends to shift toward lower voltages and the inverted U shape becomes flatter. This can be simply represented by the weak-spot ratio, which is defined as the ratio of the insulation resistance at the first-spot ratio, which is defined as the ratio of the insulation resistance at the first-step voltage to that at the second-step voltage, the two steps being specified through experience.

Generally speaking, the current decreases monotonically when steady DC voltage is applied, but kicks of current are sometimes observed in a deteriorated cable. This is probably

Table 3.3.1.
CHARACTERISTICS OF SEVERAL DEGRADATION DETECTION METHODS

Measuring method	Applied voltage	Measuring items	Characteristics
DC voltage method	DC voltage	I-t characteristics	Small size of apparatus Less damage due to overvoltage
	30 kV: 3 ~ 7 min;	Discharging current	A certain judgement standard available
	50 kV: 7 min	Partial discharges	Average in degradation of a whole cable insulation system Occasional detection of local degradation Most abundant field experience achieved Partial discharge detection is not as good as with other AC method[a]
AC method	AC 50 Hz or 60 Hz	Dielectric properties	Effective for localized defects included in a cable under service condition
		Partial discharges	Many factory test data available Long cables require large capacity electric sources[a]
AC-superposed DC voltage method		I-t characteristics Dielectric properties Partial discharges	Partial discharges can be detected more easily than by the DC method, which depend on the magnitude of AC portion of voltage DC current can be measured Less test experience[a]
Ultra-low frequency method	AC 0.1 or 0.05 Hz	Dielectric properties Partial discharges	Small charging current Partial discharge behavior per cycle appears to be equivalent to that in 50 or 60 Hz case Less damage due to overvoltage Superior in degradation detection by dielectric properties to the AC method Large size of apparatus[a]
DC voltage rising and descending method	Time constant sec ~ min	Partial discharges Discharging currents	Nearly equivalent to the AC method Measurements by the DC method are possible
Discharging voltage method	Time constant 70 msec	Partial discharges	Nearly the same with the above method A little complicated test apparatus[a]
High frequency voltage superposition method	700 Hz voltage superposed on service voltage	Unbalanced high frequency current	Expected to monitor a cable system under service condition No field testing experience[a]

[a] Disadvantages.

due to big partial discharges occurring in the cable insulation. The phenomenon can be used as a measure for judging deterioration. Extruded cables containing water trees also exhibit such kicks of current.

B. An Example of Deterioration Judgment

In spite of the apparent soundness of these methods based on DC characteristics of cable insulation, there have been few definitive references by which to judge the degree of deterioration of cable insulation. One example for 22 kV solid (oil-impregnated paper-insulated) cable is given below.

A cable is considered to be approaching failure if any one of the following conditions is fulfilled:

1. The leakage current exceeds 200 μA/km at DC voltage of 30 kV
2. The interphase imbalance coefficient exceeds 200% at a DC voltage of 30 kV
3. The leakage current exceeds 80 μA/km at a DC voltage of 30 kV and the polarity index is less than 1.0
4. The kick current corresponds to a discharge magnitude of several thousand picocoulomb

C. Partial Discharge Measurement

Partial discharges can take place in voids in cable insulation and deterioration can result. If it is the type of deterioration that increases the number and size of voids, partial discharge measurement might be used to evaluate the degree of insulation deterioration. Methods for partial discharge measurement are described in the next section.

Partial discharges are essentially absent in OF cables, POF cables, and extruded cables, thus it would be difficult to use this method for such cables. To make a judgment on 22 kV SL cables, the following recommendations should be followed.

A cable requires attention if it exhibits partial discharges

1. Larger than 1000 pC with a repetition rate of 1 pulse per second 2 to 3 min after 30 kV DC is applied
2. Larger than 1000 pC with the repetition rate of 10 pulses per second 5 min after 15 kV DC is applied
3. Larger than 2500 pC when a quasitriangular, ultra-low frequency voltage of 30 to 36 kV (peak to peak) is applied

D. Dielectric Tangent Measurement

Loss tangent measurement is widely employed as a factory test to judge or evaluate the extent of cable insulation deterioration. It measures the loss tangent of the cable insulation as a function of applied voltage and temperature. Since it is generally difficult to obtain the temperature characteristic for any installed cable system, voltage dependence is used. This provides an assessment not only by tanδ, but by Δtanδ and the ionization factor also.

The loss tangent is a very suitable measure if the insulation under investigation is fairly uniformly degraded. It may therefore be ineffective in detecting any local defects. Loss tangent measurement seems critical for extra-high voltage taped cable inasmuch as the tangent determines the power the cable can transmit.

The Schering bridge is the most popular tool for loss tangent measurement; it can be used either at power frequency (50 Hz or 60 Hz) or an ultra-low frequency. Lower frequency testing was introduced as a means of reducing the capacity of the testing power source. AC superimposed on DC may be used for the same reason.

E. A Judgment Method for XLPE Cables with Water Trees

A nondestructive method of judging the insulation condition and a scheme for lifetime estimation have been proposed for 3.3 kV and 6.6 kV XLPE cables containing water trees.[13] These are based on the results of tests of insulation characteristics and breakdown tests which were carried out on both cables removed from service and cables subjected to an acceleration test.

One way to judge the insulation quality of XLPE cables is to determine whether or not there exist water trees harmful to cable service. Harmfulness of water trees may be measured in terms of the ability of the cable insulation to support the voltage stresses required by service, but how to choose the value of the breakdown strength remains unsettled. The breakdown strength required for service is supposedly 2.5 kV/mm, which is estimated from

Table 3.3.2.
A RECOMMENDED
JUDGMENT STANDARD
ON WHETHER OR NOT
THERE EXIST WATER
TREES HARMFUL TO
SERVICE

Judgment parameter	Electrical resistance $(\Omega \cdot F)$	tan δ (%)
Judgment level	<0.01[a]	>5.0[b]

[a] 10 min values at room temperature at the same DC voltage as the AC rated voltage.
[b] Values at the AC rated voltage.

the average value of twice service voltage at 3.3 and 6.6 kV. Table 3.3.2. shows a recommended judgment standard whereby one can determine whether or not there exists water trees harmful to service.[13] Water trees are deemed to have rather a harmful level if insulation resistance is less than, and tanδ is more than, the respective critical value given in the table. Care should be exercised if either condition occurs.

Figure 3.3.1. shows a V-t characteristic for XLPE and PE cables with inner and outer semiconducting layers of cloth tapes. A data group of open marks was obtained for the short-time AC breakdown strength for some XLPE cables which had been subjected to the acceleration test and a cable removed from service. This V-t characteristic is somewhat different from the ordinary V-t characteristic which is obtained by the application of various constant voltages. The remaining lifetime can be estimated from the curve of Figure 3.3.1.

As mentioned, the figure applies to cables with cloth semiconducting layers, which are prone to form water trees. In Volume II, Section 1.1.6., we note that recent progress in cable technology makes it possible to reduce irregularities or asperities by the simultaneous extrusion of semiconducting layers with main insulation. This would make the nucleation time of water trees longer. A similar dependence may be expected, even for such better cables, once water tree formation starts.

3.3.2. Monitoring of Oil in Taped Cables

Degradation of insulation of taped cables proceeds gradually over a long period during their operation. It will be accelerated by accidental ingress of water or gases into the insulation at the times of installation and maintenance.[2] A good way to monitor insulating oil is to take it out of the cable system of interest where this is possible. Such monitoring tests can be made on selfcontained and pipe-type oil-filled cables.

Insulating oil can be collected at middle joints for three-phase conductor cables and at oil-stop joints or terminations for single conductor cables. Stringent measures should be taken when the oil is being collected to avoid its coming in contact with atmospheric air. The addition of water or oxygen that could result would cause deterioration of the oil so that its performance would be affected and false measurements would result. Since impurities other than water, such as metal, also have degrading effects, a glass container is favorable for collecting the oil. Even the glass container should be cleansed by suitable solvents and deionized water prior to test. The sampled oil must be kept away from sunlight and high temperature, and be tested as soon as possible.

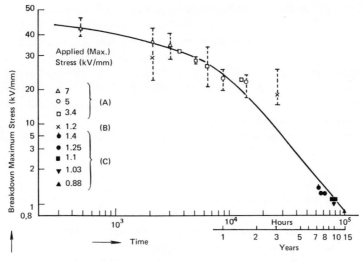

(A) *Cables subjected to an acceleration test*
(B) *Cables removed from service*
(C) *Cables to have failed due to water trees during service*

FIGURE 3.3.1. A V-t curve to estimate a remaining lifetime of XLPE and PE cables with water trees. (From Tanaka, T., Fukuda, T., and Suzuki, S., *IEEE Trans. Power Appar. Syst.*, 95(6), 1189, 1976. With permission.)

It is the present conclusion that OF and POF cables that were originally sound retain their characteristics for a long time under normal use. Accidental ingress of water or air at the time of installation and maintenance can be monitored effectively by the measurements of

1. Dielectric properties such as volume resistivity, dielectric tangent, and dielectric breakdown strengths
2. Moisture content, which is measured mostly by the famous Karl-Fischer's method
3. Dissolved-gas content

Dust, solid particles, and ionic impurities such as water may be checked by the first method. The second and third methods are suitable for detecting water content. The third method is especially suited for making a microanalysis of thermally dissolved gas which gives a clearer picture of insulation degradation. Quantitative analysis of acid value, dissolved gases, peroxides, free radicals, and cuprous ions are now under investigation as means to study the degree of insulation deterioration.

3.4. PARTIAL DISCHARGE MEASUREMENT

3.4.1. Measurement Methods

The terminology for partial discharge is controversial. Corona or corona discharge seems to be favored in the U.S., internal discharge, void discharge, and cavity discharge are other names, each of which describes a certain facet of discharge phenomena. At any rate, the partial discharge that is discussed here is a discharge or breakdown partially occurring in a dielectric, and in most cases a gaseous discharge which does not bridge the electrodes. According to the literature,[14] partial discharges fall into three classes: the internal discharge, the surface discharge, and the corona discharge, which occur in inclusions or cavities in a dielectric, at the surface of a dielectric, and in the strongly inhomogeneous field around a sharp point or edge of an electrode, respectively.

Partial discharges in the insulation of solid-type taped cables are permissible to a certain extent. No partial discharges are to be expected in selfcontained and pressure-type OF cables because their insulation structure is designed to be void-free. Extruded cable insulation is most vulnerable to failure once it is exposed to partial discharges; for this reason partial discharge measurement has been developed in order to detect gross defects in such cables which may have been produced by mistake or accident during the course of their manufacture. Much effort has been expended to detect any large detrimental imperfections produced during service, but this effort has met with limited success.

There seem to be two major purposes in further developing partial discharge measurements. They are to

1. Improve sensitivity
2. Improve the signal to noise (S/N) ratio

As noted in Volume II Section 1.1.6., one way to improve extruded cable insulation is to reduce the size and content of voids in the insulation. The size of a void subjected to partial discharges which is likely to be damaging to the cable still remains uncertain; hence the objective to develop a detector to obtain higher sensitivity. It is claimed that the measurement could be made on less than 0.01 pC, but 0.1 pC appears to be an accepted magnitude for a measurable discharge.

A. Discharge Pulse Detection

An insulation system with one void is usually represented by an equivalent circuit comprising a dischargeable capacitor C_g in series with a nondischargeable capacitor C_b; this combination being parallel by yet another capacitor, C_a.[3] Usually a coupling capacitor C_k is connected to the specimen of interest, together with an additional detecting impedance Z_d, so as to form a closed circuit for discharge pulsive currents.[14-16]

Figure 3.4.1. indicates a measuring circuit consisting of C_g, C_b, C_a, C_k, and Z_d (C_d) where C_g = capacitance of a discharge gap such as a void, C_b = total capacitance of insulation in series with the discharge gap, C_1 = specimen capacitance other than C_g and C_b, C_k = coupling capacitance, and Z_d = detecting impedance. Usually $C_k \gg C_a$. The capacitance parallel to the series combination of C_g and C_b is expressed by C_m for convenience, that is to say:

$$C_m = C_a + \frac{C_k C_d}{C_1 + C_d}$$

B. Apparent Discharge Magnitude

The total capacitance of the measuring system to which an AC voltage V_a is to be applied is given by:

$$C_a = C_m + \frac{C_g C_b}{C_g + C_b} \qquad (3.4.1.)$$

Thus the partition voltage v_g across the gap is expressed by

$$v_g = V_a \frac{C_b}{C_g + C_b} \qquad (3.4.2.)$$

When v_g reaches a critical voltage V_g, which is often referred to as Paschen's voltage in case of a dielectric with a gaseous cavity, gaseous breakdown takes place in the gap, leading to the partial or total liberation of the charge stored in the capacitor C_g prior to the breakdown. This is what is called a partial discharge of the insulation system.

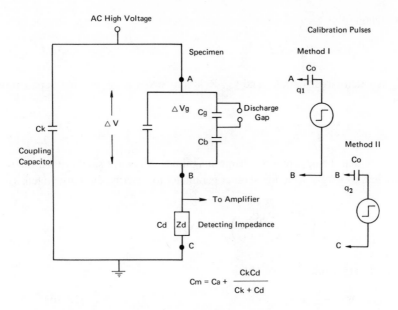

FIGURE 3.4.1. Circuit to detect partial discharge pulses.

After the discharge takes place, the voltage decreases from V_g to V_r within a quite short time, producing a current pulse $i(t)$. The latter voltage is called the residual voltage at which the discharge ceases. The duration of gaseous partial discharges is less than 1 nsec; the minimum value measured so far being 250 psec.[17] The change in the gap voltage is then given by

$$V_g - V_r = \frac{1}{C'_g} \int_0^\infty i(t)dt = \frac{q_r}{C'_g}$$

with

$$C'_g = C_g + \frac{C_m C_b}{C_m + C_b} \tag{3.4.3.}$$

The term q_r is a real charge transferred in the gap due to one discharge. We can write Formula 3.4.3.

$$q_r = (V_g - V_r) \left(C_g + \frac{C_m C_b}{C_m + C_b} \right) \tag{3.4.4.}$$

This quantity cannot in general, be measured, so it is replaced for convenience by an apparent discharge magnitude.

The apparent discharge magnitude is defined by:

$$q = \left(C_m + \frac{C_g C_b}{C_g + C_o} \right) \Delta V \tag{3.4.5.}$$

where ΔV is the voltage change on the specimen due to the discharge. The voltage change is given by:

$$\Delta V = \frac{C_b}{C_m + C_b} (V_q - V_r) \tag{3.4.6.}$$

Thus we obtain

$$q = C_b(V_q - V_r)$$ (3.4.7.)

when C_m is much larger than C_g and C_b. Relation between the real and apparent discharge magnitudes is

$$q = \frac{C_b}{C_q + C_b} \, q_r$$ (3.4.8.)

This indicates that the apparent discharge magnitude is always smaller than the real one.

The discharge energy, which is another parameter to describe discharge intensity, is given by:

$$W = \frac{1}{2} \left(C_q + \frac{C_m C_b}{C_m + C_b} \right) (V_q^2 - V_r^2)$$ (3.4.9.)

If V_g is much larger than V_r, we obtain

$$W = \frac{1}{\sqrt{2}} \, q \, V_i$$ (3.4.10.)

where V_i(rms) is the discharge inception voltage across a specimen. The last relationship indicates that the apparent discharge magnitude may be qualified as a representative parameter to describe discharge intensity rather than the real transferred charge.

C. Calibration

Apparent discharge magnitude can be obtained in principle from Equation 3.4.5. by measuring the voltage change across a specimen due to the discharge, but as a matter of practicality it is difficult to make an accurate estimate because of uncertainty in C_g, C_b, and the stray capacitances of the measuring system. Instead, it is common practice to employ calibration pulses fed by a suitable pulse generator, as is shown in Figure 3.4.1.

Pulses for calibration should have the same rise time as the actual discharge pulses. Pulses with rise times of the order of nanoseconds are employed to simulate gas discharges. Two methods are illustrated in Figure 3.4.1. Method I permits the pulses to be applied across the specimen under test. This is a direct pulse-injection method for which the injected charge q_1 is equal to q. The calibration capacitance C_0 is chosen to be much less than C_m or $C_m + C_g C_b/(C_g + C_b)$ to be exact. In this way almost all the charge can be transferred from the pulse generator to the measuring system between A and B in the figure. The charge corresponds to the calibration capacitance C_0 times the voltage pulse height of a calibration pulse generator. Pulses can be injected across the detecting capacitance (or between B and C in Figure 3.4.1.), as the second method indicates, if $C_k \gg D_c \gg C_a$, C_b, and C_g; the apparent discharge magnitude q is equal to the injected charge q_2 divided by a factor $(1 + C_d/C_k)$.

D. Measuring Circuits

The three circuits as shown in Figure 3.4.2. represent circuit systems used for discharge pulse detection. The first and second circuits are essentially the same as regards discharge pulses except that they produce pulses of opposite polarity. The first circuit requires the specimen to be insulated from ground, but does not necessarily need the choking circuit when noise from the voltage source is extremely low. The second has the specimen directly grounded, and requires the choking circuit to prevent discharge pulses from being fed back to the voltage source. When a large coupling capacitance is chosen, care must be taken for AC current flow in the detecting circuit. The balanced circuit arrangement shown in Figure 3.4.2. (C) is useful to increase the signal-to-noise ratio.

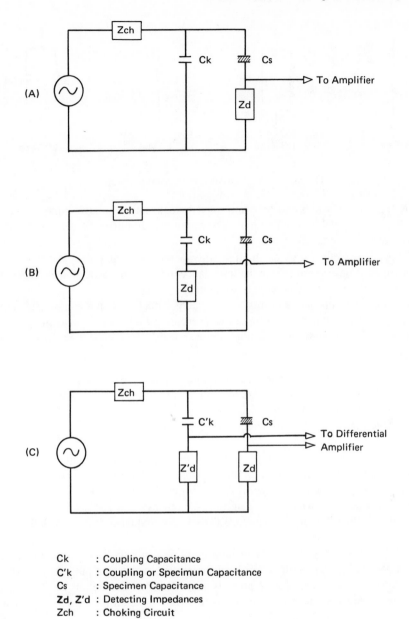

Ck   : Coupling Capacitance
C'k   : Coupling or Specimun Capacitance
Cs   : Specimen Capacitance
Zd, Z'd : Detecting Impedances
Zch   : Choking Circuit

FIGURE 3.4.2.   Three fundamental test circuits for partial discharge detection.

The detecting impedance is, of course, a combination of resistance, capacitance, and inductance, and is represented by the four major assemblies of impedances as shown in Figure 3.4.3.; R type, CR type, L type, and LC type.

1.   R type: Discharge pulse currents flowing in a specimen can be closely observed by the voltage they generate across the detecting resistance. This method therefore is suitable for investigating pulsive current response and obtaining information regarding the partial discharge mechanism. On the other hand, it is most likely to pick up various kinds of noise because of its constant frequency-response. Charging current also flows through the detecting resistance and induces appreciable AC voltage across it. To reduce the AC voltage, a high-pass filter is required.

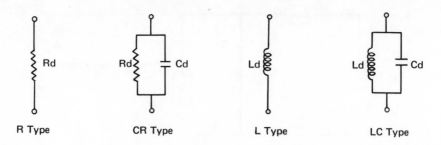

FIGURE 3.4.3. Four kinds of detecting impedance.

2. CR type: Pulse width can be determined by selecting a suitable time constant $\tau = C_d R_d$. This signal is usually fed to a wide-band amplifier (3 kHz ~ 3 MHz, for example).

3. L type: This method provides high sensitivity for discharge pulses and low sensitivity for AC charging current, since the impedance increases in proportion to the frequency. Detected pulse shape is generally oscillatory.

4. LC type: The impedance in this case is sharply tuned, but the band-tuned frequency can be adjusted. It has the advantage of preventing external noise and charging current from passing through the detecting circuit, but like the L type it produces oscillatory signals.

It is necessary to amplify detected signals in order to analyze discharge pulse trains. Three types of amplifiers for this purpose are available with different frequency responses: wide-band amplifiers, low-frequency amplifiers, and tuned-frequency amplifiers. They are equipped with a suitable high-pass filter to detect partial discharge pulses only.

1. Wide-band amplifier: This type amplifies the detected signal with high fidelity, reducing signal distortion to a minimum. Amplified pulse height is proportional to the detected pulse height. It is generally considered that this method has higher pulse resolution but somewhat lower gain than the other two. It is quite susceptible to radio noise and other external interference. The upper frequency limit is usually chosen to be of the order of megahertz.

2. Low-frequency amplifier: This method amplifies only the low frequency component of a detected pulse. In other words, it integrates the detected pulse and amplifies the integrated pulse. The resulting pulse is proportional to the area of the detected pulse. As a consequence the gain is high and free from external noise, but pulse resolution is poor.

3. Tuned-frequency amplifier: Here there are two categories, narrow band tuners, and middle band tuners. The narrow band tuner amplifies the detected pulse by a tuned amplifier with a narrow frequency band of about 10 kHz. It then demodulates or rectifies the resulting oscillatory output signal, giving its envelope as the final output. Its advantages are high gain and high S/N ratio, but pulse resolution is low since the envelope pulse is dispersed in time compared to the original signal. RIV meters belong to this narrow band-tuned amplification method. These instruments are equipped with a super-heterodyne amplifier having a frequency band of 10 kHz and a middle frequency of 455 Hz.

The middle frequency band-tuning type is an improved version of the above with respect to the pulse resolution. The frequency band is increased to 90 kHz. It should be noted that tuned devices cannot, in general, distinguish pulse polarity. Various performance data for each of the above techniques are listed for comparison in Table 3.4.1.

Table 3.4.1.
SOME CHARACTERISTICS OF THREE DIFFERENT DETECTION SYSTEMS

| Measuring method | Wide band | Low frequency | Tuning type | |
			Narrow band	Middle band
Detecting impedance	CR or LC	R and transformer coupled impedance	R paralleled by transformer coupled tuned impedance	R paralleled by transformer coupled tuned impedance
Band width of amplifier	f_1: Several kHz f_2: Order of MHz	f_1: 5 ~ 10 kHz f_2: 150 ~ 200 kHz	f_0: 200 ~ 2000 kHz Δf: 10 kHz IF: 455 kHz	f_0: 400 kHz Δf: 90 kHz
Sensitivity	Moderate	High	High	Moderate
Pulse resolution	0.1 ~ 10 µsec	20 ~ 30 µsec	200 µsec	25 µsec
Display device	Pulse counter Oscilloscope	Pulse counter Oscilloscope Volt meter	Volt meter Oscilloscope	Pulse counter Oscilloscope Volt meter
Basic quantities of measurement	Vi, q, n pulse shape	Vi, q, n pulse polarity	Vi, q RIV value	Vi, q, n
Noise Easy to pick up	Radio noise Burst noise	Burst noise	Burst noise	Burst noise

Besides the amplifiers described above, several other components are necessary to form a measuring system. The various types of partial discharge measurement systems consist of four basic subsystems: a pulse amplifier, a pulse counter, a pulse height discriminator, and a cathode-ray oscilloscope. A pulse-peak meter and an average current meter are both convenient to record the maximum discharge magnitude within a certain interval and the discharge current, respectively.

E. Measurement of Partial Discharges in Cable

There are two cases to consider as regards detecting partial discharges in cable. The first corresponds to a short cable specimen, constituting a circuit with lumped parameters. The second corresponds to a long cable such as a reeled cable or an installed cable. Here, the lumped parameter representation is inadequate; it must be represented by a circuit with distributed constants. The former can be handled comparatively easily with the principles mentioned so far, but the latter is more difficult to treat and needs further consideration.

In a factory test, a long cable can be sectionalized electrically so that it may be treated as a lumped parameter circuit. This is called the discharge scanning technique. There are two types available according to whether high voltage is applied to the insulation screen (or equivalent) or to the conductor. Figure 3.4.4. illustrates a bridge circuit for continuously scanning unscreened cable core, which constitutes "two" high voltage electrodes.[16] Single electrode scanners are also used.

Discharge scanning can be conducted on a cable with a screen if high voltage is applied to its conductor. This technique has two major disadvantages: the entire length of cable must be energized for the duration of the test, thus necessitating a high kVA rating transformer, and charging current flowing in the cable screen can cause damage to the screen.

From the viewpoint of a pulse circuit, a long length of power cable is equivalent to a distributed-constant circuit with a certain capacitances-to-ground (farad/meter), and a series of inductances (henry/meter), which contribute to the surge impedance $Z_0 \simeq \sqrt{L/C}$ (Ω). Pulses propagate through a cable conductor with a velocity of $v = 1/\sqrt{LC}$ (meter/second). The equivalent circuit shown in Figure 3.4.5. is appropriate in these circumstances.

When the voltage across the discharge gap starts to decrease due to a discharge, with a

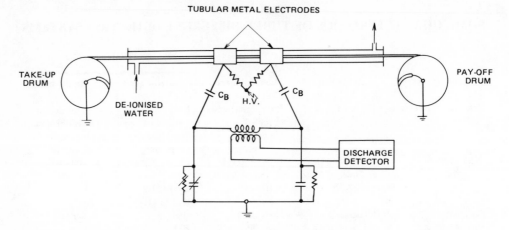

FIGURE 3.4.4. Bridge circuit for continuous scanning of cable core. (From Arnold, A., An Introduction to Partial Discharge Detection in Cables, Lecture Note, Brighton College of Technology, Brighton, England, p. 11.)

time constant τ_p, a pulse voltage V(t) which is given by

$$V(t) = \frac{q(Z_0/2)}{\tau_p - \tau_0} \left\{ \exp \left(-\frac{t}{\tau_p} \right) - \exp \left(-\frac{t}{\tau_0} \right) \right\}$$

with

$$\tau_0 = C_b Z_0/2 \qquad\qquad (3.4.11.)$$

is induced at the discharge point. Since Z_0 and C_b are 20 Ω and several pF, respectively, τ_0 is of the order of 10^{-11} sec, which is small compared with τ_0 (order of 10^{-9} sec). Accordingly, the voltage shape is determined almost entirely by the discharge time constant.

When a detection circuit is added to a cable system in such a manner that the detection resistance R_d is connected to its detecting terminal through a coupling capacitor as is depicted in Figure 3.4.5., part of propagating pulse is absorbed. A single discharge pulse produces a number of pulses at the detection point due to pulse reflection at the two terminals. When the time constant $C_k R_d$ is large enough, a train of pulses can be observed, the first pulse of which is $\{2R_d/(R_d + Z_0)\}$ times the original signal.

The wide-frequency band method produces a similar pulse train to that just described. The first pulse of the train gives the discharge magnitude. In principle the discharge point can be located from the time difference between the first and subsequent pulses. The superposition of time-spaced multiple discharge pulses is likely to occur when the low-frequency band method is used because of its integrating characteristic. However, it may be possible to resolve the pulse train if cable under test is more than several hundred meters long, thereby making it possible to determine the discharge magnitude from the first pulse. The tuning-type method produces oscillatory signals, which are then integrated. Different output is obtained according to the polarity of signals reaching the detection terminal. In other words, the detection sensitivity of this method changes with cable length. Detection sensitivity as a function of cable length for the three methods is indicated schematically in Figure 3.4.6. A differential bridge-type detection circuit is preferable in order to increase the S/N ratio.

3.4.2. Partial Discharge Test Specification

Partial discharge tests for extruded cables are specified in AEIC No. 5-75.[6] The test apparatus and calibration procedure is described in IPCEA Publication T-24-380, 1974, *Guide for Test Procedure*.[18]

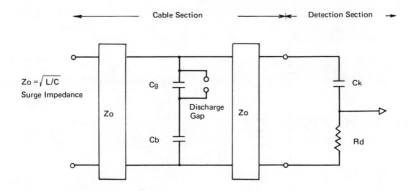

FIGURE 3.4.5. Equivalent circuit for partial discharges occurring in a long cable and detection circuit.

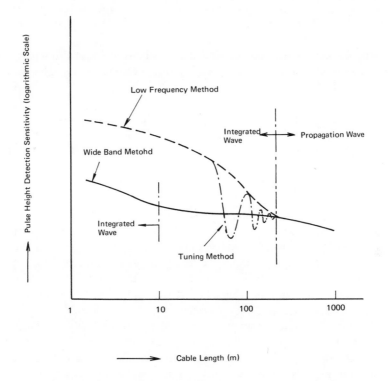

FIGURE 3.4.6. Variation of detection sensitivity with cable length.

An alternating current test voltage, having a frequency within the range 49 to 61 Hz, shall be applied between the conductor and the metallic component of the insulation shield of the individual conductors under test. The rate of change of the applied voltage shall not exceed 2000 V/sec as the voltage is decreased. The test voltage shall not be maintained for more than 3 min during any single test. The applied test voltage shall be raised to the appropriate maximum voltage level given in Table 3.2.4. and then lowered in order to derive the apparent discharge characteristic. If the apparent discharge magnitude exceeds 5 pC at the maximum test voltage, the extinction shall be noted as the applied voltage is being lowered. The cable under test shall be so terminated that there will be no detectable partial discharge on the ends of the cable over the range of partial discharge test voltages required for that cable.

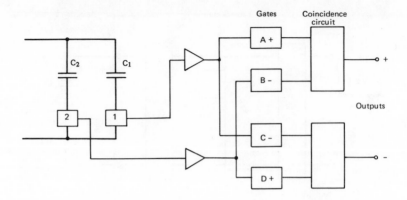

FIGURE 3.4.7. "Digital Bridge" circuit to reject external noise in partial discharge measurement. (From Black, I. A., *A Pulse Discrimination System for Discharge Detection Measurements on Equipment Operating in a Power System,* IEEE Conf. Publ. 94, Part 1, Piscataway, N. J., 1973, 5. With permission.)

A two-purpose generator is recommended to calibrate the partial-discharge measuring device for pulse resolution time and and superposition characteristics. The pulse generator shall have these characteristics:

1. Produce identical variably time-spaced 40 pC discharges (pulses)
2. Time spacing variable from 1 to 100 μsec
3. Maximum rise time of pulses — 20 nsec (rise time from 10% of peak value to 90% of peak value)
4. Output impedance — 1400 Ω ± 20%.

3.4.3. Advances in Partial Discharge Measurement

As mentioned in Section 3.4.1., the areas for further development of partial discharge measuring equipment are sensitivity and S/N ratio, both of which one would like to increase. As far as sensitivity is concerned, it may be hard to improve on the performance of selected amplifiers as described. A higher signal-to-noise ratio may be obtained by using a balanced bridge circuit and by testing in a shielded room if possible. Suitable logic circuitry in the balanced type of circuit appears to increase the S/N ratio considerably.

The application of a multichannel pulse height analyzer, originally developed in the field of nuclear spectroscopy, and the application of a versatile minicomputer, are other avenues for improved performance. Many parameters are available to describe partial discharge phenomena. Unfortunately, no comprehensive study has been made which embraces all the parameters, simply because it is almost impossible to measure and process all the partial discharge data with any conventional technique. It can be done by a minicomputer system. Indeed a multichannel pulse height analyzer alone is a great help.

A. Logic Circuitry for Increasing the S/N Ratio

This essence of this approach is to reject noise by checking the polarity of two signals appearing across two detecting impedances in a balanced bridge-type circuit. Figure 3.4.7. shows a logic circuit which provides this function.[19] Discharges external to C_1 and C_2 produce co-directional current pulses in the capacitors, whereas discharges within C_1 and C_2 produce opposing currents. The input units in series with C_1 and C_2 pick up voltage pulses due to discharges and feed them, via amplifiers, to the gates as shown. If both gates operate simultaneously, the coincidence circuit is triggered and gives an output pulse. Since the gates are polarity-sensitive, they can be arranged to trigger with opposite polarity pulses and thus indicate the presence of discharges in either C_1 or C_2. External pulses will not trigger the circuit under these conditions. A partial discharge measuring system equipped with a logic circuit of this kind was reported to exhibit a rejection ratio of 7000:1.[19]

B. Signal Recovery Technique — Correlation Analysis

Partial discharge pulses appear synchronously in phase with the applied AC voltage, while radio noise or spurious transients are out of phase in most cases. A technique which relies on signal correlation may be effective in such cases. It must be remembered, however, that a discharge of a certain magnitude does not necessarily occur at a specific point on the voltage cycle. There is a limit to the application of correlation techniques to signal recovery, the degree of success being dependent on the degree of repeatability of the discharge sequence each cycle.

The signal train from an amplifier is divided into increments corresponding to 1 cycle (20 msec for 50 Hz voltage and 16.7 msec for 60 Hz voltage) and then into 100 or 200 subincrements. The signal level is then stored and averaged with the level in the same subincrement of the next period, and so on, for say, 1000 successive cycles. The final average for all subincrements in the 1 cycle time (20 msec or 16.7 msec) is then stored and displayed. Random effects are averaged to zero and the output represents the probability of a discharge of a given size during each particular subincrement. Since discharges tend to occur on the two rising voltage quadrants, subincrements in these regions have an output while those relating to other parts of the waveform do not. Repeating the test at a higher voltage and comparing the averages at various levels permits the determination of inception voltage and charge magnitude at inception and working voltages. A correlator with such a function is commercially available. This system combined with a suitable amplifier (a narrow band 500 Hz or 15 to 150 kHz unit) was reported to have detected 1 pC against a background noise level equivalent to 60 pC, and 0.2 pC for a 5 pC noise background.

C. Application of a Multichannel Pulse Height Analyzer

With the use of a multichannel analyzer, entire spectra of partial discharge pulses may be examined with relative ease to obtain statistical probability data on the distribution of the discharge pulse recurrence rates as a function of the discrete pulse amplitudes. A partial discharge measuring system with such a multichannel pulse height analyzer is illustrated in Figure 3.4.8., with a cable shown as the test specimen. Data are displayed in the form of the pulse repetition vs. pulse magnitude on an oscilloscope (or an XY recorder). The system depicted in Figure 3.4.8., is of the category using the undamped response signal detection mode; its basic sensitivity level is of the order of 1 pC, and the pulse resolution limit of the pulse-shaping circuitry itself is in the range 10 to 15 μsec. It is capable of further improvement. An assembly of both a pulse height analyzer and a display unit, computer-controllable, is commercially available. Such systems are also known as monitored corona spectroscopes.[22]

A pulse height analyzer in general allows only unidirectional pulses (positive or negative). But it is possible to measure both polarities if the memory device of a pulse height analyzer can be externally controlled. A system consisting of an amplifier, an inverter, and a mixer and router unit, coupled with the pulse height analyzer, can store two kinds of pulse trains in two different regions of the memory device. This might be positive pulse trains and inverted-negative (actually positive) pulse trains.

D. Computer-Aided Partial Discharge Measuring and Data Processing System

This system, in which versatile functions of a minicomputer such as memory, data processing, and judgment are fully utilized, has been developed in order to make comprehensive studies of partial discharge phenomena. It is designed to handle the magnitude distribution characteristics of partial discharges and also as a transient recorder to analyze wave shape. Measurement and data processing can be conducted with the minicomputer in the conversational mode via a teletypwriter.

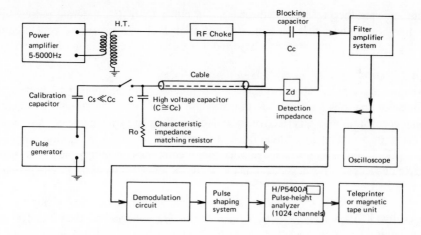

FIGURE 3.4.8. Partial discharge measurement system using a pulse height analyzer. (From Bartnikas, R., Use of a multichannel analyzer for corona pulse height distribution measurements on cable and other electrical apparatus, *IEEE Trans. Instrum. Meas.*, 22(4), 404, 1973. With permission.)

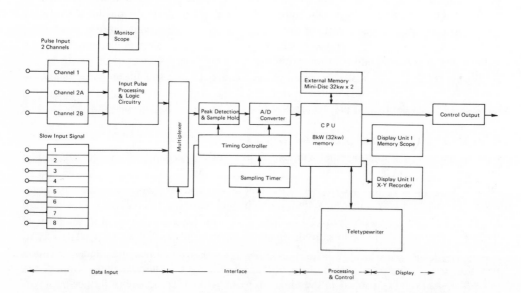

FIGURE 3.4.9. Block diagram of partial discharge measurement and data processing system.

Figure 3.4.9. shows a block diagram of this system. It will be seen to comprise a data input section, an interface section, a data processing and control section, and a display section, together with an appropriate computer program as will be explained later. Since this system tends to be software oriented with the suitable hardware, it is flexible enough to adapt to standard and nonstandard data processing requirements. By a control output from the central processing unit (CPU), this system can be extended to control a measuring circuit, thereby producing a real-time, on-line, partial discharge measurement system, a real advance in discharge measurements. This can be expanded further for cable fault location.

There are three basic data input methods for analyzing discharge pulse trains.

1. Pulse height distribution
2. Pulse height in time and phase domains
3. Pulse repetition rate in time and phase domains

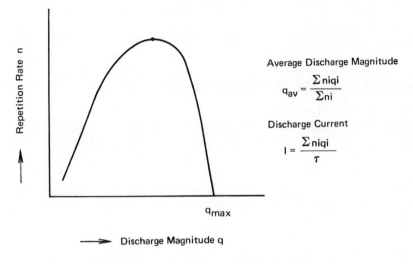

FIGURE 3.4.10.   Discharge magnitude distribution.

The first method functions in a way similar to a commercially available pulse height analyzer except that this system accepts both positive and negative pulses, which are measured simultaneously. Data taken into the CPU are processed and displayed as

1. Separate graphical and digital displays of repetition rate vs. discharge magnitude for both positive and negative polarities
2. A digital display of representative quantities in pulse height distribution as shown in Figure 3.4.10.; the maximum discharge magnitude q_{max}, the most frequent discharge pulses (q_{peak}, n_{peak}), the average discharge magnitude $q_{av} = (\Sigma n_i q_i)/(\Sigma n_i)$, and the discharge current $I = (\Sigma n_i q_i)/\tau$
c. The difference between two pulse height distributions (this can be utilized to increase the S/N ratio)
d. A correlation and a deviation in an assembly of pulse height distributions

The second input method is concerned with discharge pulse magnitudes in time or phase domain. Both positive and negative pulse trains can be processed by averaging pulses appearing in the time phase; the technique is similar to that already described for signal recovery. It gives mean values, root mean square values, and standard deviations. These can be displayed graphically and digitally. This resolves temporal characteristics of pulse trains and as a result increases the S/N ratio. Time-spaced characteristics of discharge pulses obtained by this method can be transformed and replotted as repetition rate vs. discharge pulse height.

The third method concerns repetition rate for discharge pulses which exceed a given, adjustable level. The data processing is exactly the same as that used in the second method except that pulse repetition rate replaces pulse height.

In addition to the three basic functions, long-term measurement of representative quantities and voltage dependence of discharge pulse height are available.

This is a new and promising approach for clarifying phenomenological aspects of partial discharge pulse behavior with respect to pulse groups. Such a system is described in Reference 23.

3.5. FAULT LOCATING

All types of cable are subject to electrical failure. It is important to locate the point of fault as soon as possible. OF, POF, and GF cables can experience other types of failure which may be followed by electrical failure. Oil leakage accidents for OF and POF cables and gas leakage accidents for GF cables are examples.

3.5.1. Detection of Oil Leaks

The location of an oil-leak point can be established by visual inspection, oil-flow behaviors, or by frozen cable methods.[2] It is possible to use a semistop joint to sectionalize a cable; this seems to be effective for gross oil leaks. A thermal flow method may be useful for a POF cable. This involves judging the oil flow direction from axial variation of temperature on the pipe when part of the pipe is heated. When the oil leak is caused by external forces, the protective covering of the cable is damaged and it may also be earthed. In this event it may be possible to utilize a ground point detection technique.

A. *Visual Inspection*

The first thing to do after it has been established that oil is leaking from a cable is to inspect any suspect and accessible point of the line, such as existing or former construction sites, places where the ground is susceptible to subsidence, exposed parts of the line, termination joint boxes, straight joint boxes, and oil-supplying systems. If there is stagnant water in manholes and elsewhere, floating oil on the water can be checked visually and by suitable chemicals.

B. *Making Use of Oil Flow Characteristics*

The way oil flows in OF cable can be utilized to pinpoint oil leaks. Several methods are listed with comments in Table 3.5.1.[2] Expansion and contraction of oil due to temperature change may upset the location sensitivity when the cable of interest has a small oil leak or a wide temperature excursion. In such cases, another sound, parallel cable line is used to compensate for this effect.

It takes time for oil flow to reach a steady state in a cable. The differential pressure method II, the bridge method, and the pressure drop method, all require a change in direction of the oil flow. Comparison methods I and II ensure equi-pressure of oil in the system under test, while differential pressure method II, the bridge method, and the pressure drop method assume equi-flow of oil.

The oil pressure method requires a correction for hydrostatic pressure arising from differences of level in the installation site. It is therefore inappropriate for detecting small oil leaks. The oil-flow method II takes advantage of the fact that immediately after the oil supply valve has been closed, the pressure difference between the oil feed bath and the oil feed inlet is almost inversely proportional to the distance to the point of any oil leak. The standard oil-flow resistance is obtained by means of a pipe with a homogeneous inner diameter, placed in a constant-temperature bath to keep the oil viscosity constant.

C. *Frozen Cable Method*

The cable of interest can be divided in two by freezing it at a certain point. One can then judge on which side the oil leak is located. The leak can be localized by repeating the procedure. The oil feed should be sufficient to maintain an acceptable pressure level. Liquid nitrogen is used as the freezing agent for OF cables and POF cables with synthetic pipe oil, whereas a mixture of methanol and dry ice (solid carbon dioxide) is used for POF cables with mineral pipe oil.

For POF cables, the freezing time can be obtained from the following empirical formula:

$$T = K(D/L)^2 \qquad\qquad (3.5.1.)$$

Table 3.5.1.
VARIOUS METHODS FOR DETECTING OIL LEAK POINTS

Method	Measurement Method	Calculation Formula	Number of Cables for Measurement	Effect of Temperature	Number of Measurements	Number of Measuring Spots	Oil Feeding?
Oil Pressure Method	Pressure should be corrected by hydrostatic pressure	$x = \dfrac{P_A' - P_C'}{P_A' - P_B'} \times \overline{AB}$	1	Large	1	Many	NO
Oil Flow Method I		$x = \dfrac{Q_2 - Q_1}{Q_2 + Q_3 - 2Q_1} \times 2\ell$	3	Small	1	1	NO
Oil Flow Method II	Measure pressure just after valves are closed	$x = \dfrac{\Delta P_2 - \Delta P_1}{\Delta P_2 + \Delta P_3 - 2\Delta P_1} \times 2\ell$	3	Small	1	1	NO
Differential Pressure Method I	$\Delta P_D = \Delta P_C$	$x = \dfrac{\Delta P_C}{\Delta P_B} \times \overline{AB}$	2	Small	1	Many	NO
Differential Pressure Method II	ΔP_S with V_A open & V_B closed, and ΔP_A with V_A closed & V_B open	$x = \dfrac{\Delta P_B}{\Delta P_A + \Delta P_B}\ell$	2	Intermediate	2	2	YES
Differential Pressure Method III	V_1 open & V_2 closed; V_1 closed & V_2 open	$x = \dfrac{\Delta P_1 - \Delta P_2}{\Delta P_1}\ell$; $x = \dfrac{2\Delta P_1 - \Delta P_2}{\Delta P_1}\ell$	3	Small	1	1	NO
Bridge Method I	ℓ_c: equivalent cable length	$x = \dfrac{\Delta P_2}{\Delta P_1 + \Delta P_2}\ell$	2	Intermediate	2	2	YES
Bridge Method II	R_2 adjusted so that $\Delta P = 0$	$x = \dfrac{R_2}{R_1 + R_2} \times 2\ell$	2	Small	1	1	NO
Pressure Drop Method	ΔP_1 with V_1 open & V_2 closed, and ΔP_2 with V_1 closed & V_2 open	$x = \dfrac{\Delta P_2}{\Delta P_1 + \Delta P_2} \times 2\ell$	2	Intermediate	2	1	NO
Comparison Method I		$x = \dfrac{\Delta P_1 - \Delta P_2}{\Delta P_1 + \Delta P_2}\ell_c + \dfrac{\Delta P_2}{\Delta P_1 + \Delta P_2}\ell$	1	Large	1	2	YES
Comparison Method II	ΔP_2 with V_{A1}, V_{B2} open & V_{A2}, V_{B1} closed and ΔP_1 with V_{A1}, V_{B2} closed & V_{A2}, V_{B1} open	$x = \dfrac{\Delta P_1 - \Delta P_2}{\Delta P_1 + \Delta P_2}\ell_c + \dfrac{\Delta P_1}{\Delta P_1 + \Delta P_2}\ell$	2	Small	1	2	YES
Comparison Method III		$x = \dfrac{\Delta P_1 - \Delta P_2}{\Delta P_1 + \Delta P_2}\ell_c - \dfrac{\Delta P_1}{\Delta P_1 + \Delta P_2} \times 2\ell$	3	Small	1	1	NO

Symbol	Explanation	Symbol	Explanation
(PT)	Pressurized Oil Reservoir	ℓ	Cable Length
V ⊗	Valve	x	Distance to A Point of Oil Leak
Q	Flow Meter	Q	Flow Quantity
P	Pressure Gauge	P	Pressure
(ΔP)	Differential Pressure Gauge	ΔP	Differential Pressure
R ⌁	Standard Oil Flow Resistance	R	Standard Oil Flow Resistance
⌇	A Point of Oil Leak	ℓ_c	Cable-Equivalent Length of Rc

From Iizuka, K., **Ed.,** *Power Cable Technology Handbook,* Denki-Shoin, Tokyo, 1974, 587. With permission.

where L = the length of a freezing box (mm), D = the diameter of the cable pipe (mm), and T = the time required for freezing.[24] K values of 120 and 160 have been reported for mineral oil and alkylbenzene oil, respectively, from experiments on 275 kV, 2 × 1400 mm² POF cables (oil flow rate 20ℓ/hr, liquid nitrogen refrigerant, D = 250 mm, L = 1000 ~ 1500 mm). As the oil flow rate or the differential pressure across the frozen portion is increased, more time is required for complete freezing until ultimately it is practically impossible to freeze a cable section. While it is frozen, care must be taken not to subject a cable to vibration, mechanical shock, or bending. Since voids are generated in insulation layers after thawing, it is necessary to apply sufficient pressure to the cable for a certain time before it is returned to service.

3.5.2. Detection of Gas Leaks
There are several methods available to detect gas leaks in gas-filled cables.

1. Bubble observation method, applicable for termination joint boxes, straight joint boxes and other cable accessories
2. Gas flow method, applicable for cables
3. Tracer gas method
4. Acoustic detection method

A. Bubble Observation Method

The application of a soapy liquid with a brush will help to detect a gas leak by checking gas evolution. Immersing a suspect section into water is also useful. In this case, the addition of about 0.1% silicone-resin bubble extinguishing agent to the water will decrease its surface tension by 50%, thereby facilitating bubble evolution from any leak.

B. Gas Flow Method

This method estimates the position of a gas leak by measuring the pressure gradient caused by the leak. It is less effective for detecting a small gas leak, but it is simple and appropriate for making a first cut at the location of a leak. A number of alternatives have been developed:

Two section method — Pressure is applied to one end of the cable of interest. After the steady state gas pressure has been reached, pressures are measured by a mercury manometer at both ends of the cable and at several points between. Plots of measured pressures vs. distance may permit the location of a leak by interpolation.

Equi-pressure method — A pressure P_A is first applied to one end of a cable and the pressure P_L is then measured at the other end. The same procedure is then followed but in reverse order, except that the pressure P_B at the second end is adjusted so as to make the pressure at the first and equal to the pressure P_L of the first test. The distance from the first end to a gas leak is estimated from the following expression:

$$x = \frac{(P_A - P_L)L}{(P_A - P_L) + (P_B - P_L)} \qquad (3.5.2.)$$

where L is the length of cable under test.

Differential pressure method — When there is a sound cable parallel with the faulted line, it is possible to locate a fault by measuring pressure differences between the two cables at several joint points. In case of three-core, low-pressure, gas compression cables, a gas-tight tube enclosed within the cable can be used for the same purpose.

Triple-manometer method — Triple manometers capable of measuring differential pressure between three points are set up at every two successive joints of a three-core, low pressure, gas compression cable.

C. Tracer Gas Method

In this method, a certain amount of a tracer gas is injected into the cable. Conditions are then monitored by a suitable detector. Halogenated gases or radioactive isotopes are available; the former are in common use because they are easily handled. Freon® and SF_6 are recommended as tracer gases because they have good insulating properties, do not damage cable insulation or metal sheaths, are chemically stable, and are readily available.

Halogenated gas detectors are commercially available. They depend for their operation on the fact that cations are generated in coaxial platinum electrodes with a heated anode, the number increasing in the presence of an halogenated gas mixture. The recommended fraction tracer gas to host nitrogen is several percent; this is a compromise based upon detection sensitivity and the probable ill effects of the tracer gas on the cable insulation.

D. Acoustic Detection Method

This method depends for its success on the detection of acoustic waves which accompany the leakage of pressurized gas. Supersonic waves generated by collisions between gas molecules at the leak point are generally measured. The fact that the detector must be in direct contact with the surface of a cable restricts the use of this method.

E. The Canine Method

In the U.S., dogs have been trained to detect oil leaks by the odor of the leaking oil. The method is similar to that applied by customs officers in the detection of contraband drugs.

3.5.3. Cable Fault Location

In general, the location of a cable fault is accomplished in three different steps: (1) recognition of a fault condition by relay operations and checks of insulation resistance and conductor resistance, (2) estimation of the point of failure by suitable methods, and (3) confirmation of the point of failure. Fault-locating methods in present use fall into two categories: terminal measurements and tracer methods.

A. Recognition of a Fault Condition

Fault conditions can be understood to a certain degree from the behavior of relays. Succeeding tests on insulation resistance and conductor resistance will provide more precise information concerning which phase of the cable has been faulted, the value of the insulation resistance between the cable cores and between the faulted cable and ground, and whether or not the cable is disconnected or is short circuited.

B. Terminal Measurement Methods

Terminal measurement methods are those which involve measuring an electrical characteristic of the faulted conductor from one end of the cable and comparing this with the same measurement made under unfaulted conditions, in terms of distance to the fault.[1] Four techniques are available for such measurements: (1) Murray loop method, (2) reflection methods, (3) discharge-type pulse radar method, and (4) capacitance measurement method.

C. Murray Loop Resistance Bridge Method

This method is based on the principle of the Wheatstone bridge. It will give the distance to the point of failure when the bridge is balanced in the manner shown in Figure 3.5.1. (A). One form of the Murray loop employs a few hundred volts only. Another operates at several kilovolts.

The Murray loop method is favored because of the high accuracy of measured values (less than 1% error), its easy manipulation when taking measurements, and the handiness of the instrument. This method is effective where either the bridge voltage is high enough or the parallel fault resistance is low enough to obtain cable length sensitivity. It is ineffective

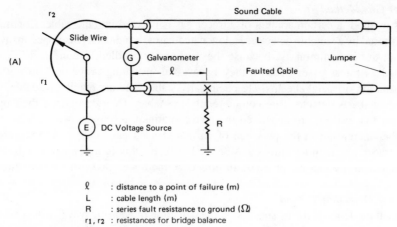

ℓ : distance to a point of failure (m)
L   : cable length (m)
R   : series fault resistance to ground (Ω)
r_1, r_2 : resistances for bridge balance

$$\ell = \frac{2r_1}{r_1 + r_2} \cdot L$$

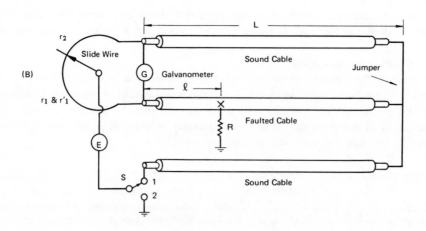

r_1 : balanced resistance when S is positioned to 1.
r'_1 : balanced ersistance when S is positioned to 2.

FIGURE 3.5.1.   (A)The Murray Loop Bridge method and (B) the Murray-Fischer method
for fault location.

for series faults of any kind. The method is particularly useful for single line-to-ground
faults which are the most frequent. On the other hand, it cannot be applied to either an open
conductor fault or to a simultaneous fault on all three phases where no parallel sound cable
remains. When the fault resistance is high (of the order of Megohms), the high-voltage
method is employed. It is impossible to make a measurement of this kind when an arc is
generated at the point of default. The application of a high voltage to a cable with a high-
resistance fault can burn out the fault and reduce the resistance to such a value as to permit
the effective use of the Murray loop method.

   The Murray-Fischer method is effective when no parallel sound cable is available or when
a defect in the corrosion protective coverings is to be detected. The principle of the method
is illustrated in Figure 3.5.1.(B). The return line indicated in the figure is not necessary
when the fault resistance is seveal tens of kilohms or higher. The far end of the cable can
be grounded when the Murray loop circuit is employed.

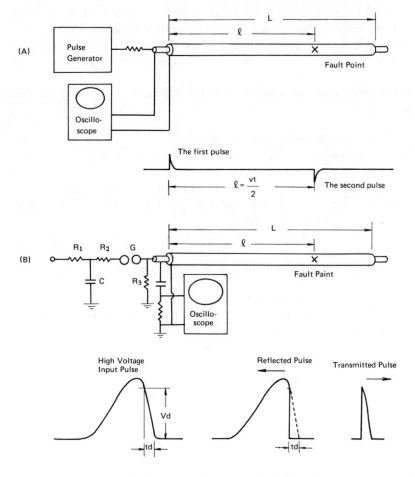

Vd : Voltage enough to cause discharges, resulting in apparent low resistance fault.
td : Delay tmie for reflected pulse, resulting in error.

FIGURE 3.5.2. (A) Low voltage pulse radar method and (B) high voltage pulse radar method for fault location.

D. Reflection Methods

Reflection methods are effective when either the parallel fault resistance is low enough, or the series fault resistance is high enough, to produce a pulse or standing wave reflection which can be distinguished from those caused by other cable discontinuity characteristics. They are ineffective on faults having high parallel or low series resistance.[1] They may be further subdivided according to technique into the pulse echo or radar method and the standing-wave or one quarter wave resonance method. In this respect, they can be called the time domain or frequency domain methods, respectively. In general, a monitoring signal is injected at one end of the cable of interest. However, there is a variant of the pulse radar method which depends on the occurrence of discharge pulses at the point of failure when a DC voltage is applied. This is referred to as the discharge-type pulse radar method.

The regular pulse radar method impresses a voltage pulse from a generator on the faulted cable and measures the time required for the pulse to reach the fault and reflect back to the source, the travel time being an indication of the distance to the fault. The basic measuring circuit is shown in Figure 3.5.2. One approach uses a pulse voltage of less than 500 V with 0.2 to 5 μsec duration; another employs a pulse voltage of several kV with a width of the order of 10 μsec. The latter is applicable when the fault resistance is high; it actually reduces the resistance by the discharges which occur at the fault when the high-voltage pulses are applied. Experimental values for the surge propagation velocity for polyethylene-insulated cables and oil-impregnated paper-insulated cables, respectively, of 172 m/μsec and 158 m/μsec have been measured. Their respective theoretical values are 197 m/μsec and 160 m/μsec.

It must be pointed out that a straight joint has a different surge impedance from that of the cable itself and therefore reflections are initiated when such joints occur. A reflection at the point of failure of at least 10% is considered to be necessary in order to distinguish the signals from such disturbing reflected pulses.

The pulse radar method is characterized by a number of advantages:

a. It is applicable to a variety of fault conditions including ground faults, open conductor faults, and short-circuited faults, and it is especially effective for simultaneous faults involving ground and open circuit.
b. It is suitable for three-phase simultaneous faults, since the measurement requires no parallel sound cable.
c. It is applicable even when the length of cable is uncertain.

On the other hand, it is not especially accurate in location sensitivity and it requires skillful operators. Errors can arise when the high voltage pulse radar method is applied to a high resistance fault because of the time lag of discharges.

There is yet another type of the pulse radar method available, the so-called differential radar pulse method. It injects pulses into both faulted and sound phases of a cable line simultaneously, and measures echo pulses differentially. In this way any pulses reflecting from points other than the faulted point are reduced in magnitude.

E. Modified Pulse Radar Measurement Method

As mentioned already, the low-voltage pulse radar method is not effective for faults with a high-resistance path to ground and with the conductor continuing past the point of the fault, conditions which are most likely to prevail on high-voltage cable. On the other hand, the high-voltage pulse radar method, which is applicable to such cases, is not very accurate. A combined system has been proposed[25] which includes all the basic features of a low-voltage system with the addition of a high-voltage pulse generator and a time-interval counter having an accuracy of 1 nanosec.

When the high voltage pulse (say 15 kV, 100 A) is applied to the cable, a discharge or an arc is formed at the fault, which appears as a short to the low-voltage radar pulse. Thus the high-impedance fault is reduced to a nearly dead short for the brief period of the discharge. During this time the low-voltage radar pulse is applied to the cable. A conceptual schematic of the system is shown in Figure 3.5.3.

F. Discharge Detection-type Pulse Radar Method

Generation of a discharge pulse at a point of failure by the application of high-voltage DC or impulse voltage gives information on fault location as indicated in Figure 3.5.4. When the fault under investigation has a high-resistance path to ground, a DC voltage is applied in order to cause a discharge at the fault. A rectangular pulse, equal in magnitude but opposite in polarity to the DC voltage, is generated in the fault and propagates in both directions toward the measuring end and the remote end. The pulse directed to the measuring terminal is recognized and measured as the first pulse. At the same time it reflects at this end and returns to the point of failure where the discharge still persists. The pulse therefore reflects back yet again toward the measuring end and is recognized as the second pulse. The time interval between the first and second pulses determines the distance to the fault. The first pulse directed toward the far end will exhibit exactly the same behavior in the

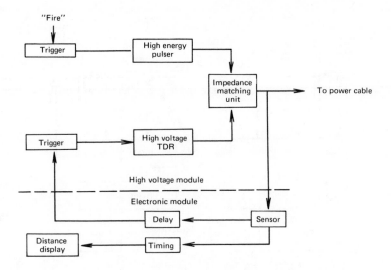

FIGURE 3.5.3.   A pulse radar measurement system with a time domain reflectometry plus a high voltage pulse generator and a time-interval counter. (From Genschke, G. P., May, K. P., and Thomas, J. A., Fault locator speeds restoration, *Electr. World*, January 1975, p. 44. With permission.)

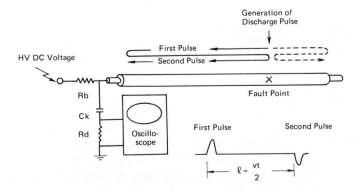

FIGURE 3.5.4.   Discharge detection type pulse radar method.

region beyond the fault, and will not interfere with signals at the measuring end. The resistance of fault remains low due to the continuing discharges; it effectively decouples the two parts of the cable. In case of a low-resistance path to ground, an impulse voltage is used to create the discharges at the fault for the required interval.

This method has the following advantages:

a.   It is applicable to high-resistance ground faults as well as low-resistance faults, short circuits, and faults due to breaks in the conductor. It is not necessary to burn a high-resistance ground fault in order to reduce the resistance.

b.   It is uninfluenced by the discharge time lag and has a measuring accuracy of less than 2 ~ 5%.

c.   It can be applied where the length of the cable is uncertain.

d.   It is effective on three-phase simultaneous faults.

There are, however, some disadvantages:

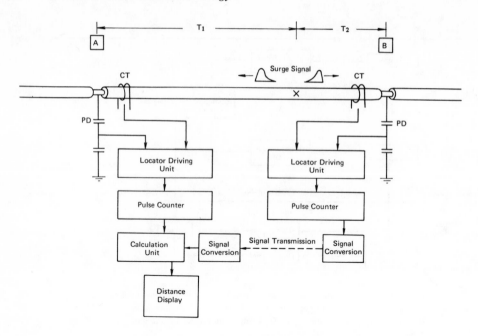

FIGURE 3.5.5.   Automatic fault locating system.

a.   The instrument required is rather large.
b.   In accuracy it is somewhat inferior to the Murray loop method.
c.   It requires skill in manipulating the instrument and reading the pulse shape.

The frequency domain approach is based on quarter wavelength resonance and the principle of wave reflections.[1] When an electrical wave of sinusoidal shape propagating along a cable encounters a discontinuity such as a short circuit (parallel fault of very low resistance) or an open circuit (series fault of very high resistance), it initiates a reflected current wave. This reflected wave, which travels back to the sending end, is only slightly diminished in magnitude and at certain critical frequencies (depending on line length) the reflected current arrives back 180° out of phase, thereby giving the cable the appearance of a high impedance or open circuit. The advantages of the one quarter wave resonance technique over the other reflection methods is that the distance to the fault appears only as a function of the dielectric constant of the cable insulation and the frequency of the current. Thus, variations in conductor size are less troublesome. However, differences in insulating materials, which may affect the dielectric constant, can introduce an appreciable error in calculating the distance to the fault.[1]

If economically acceptable, it is convenient to provide a cable system with an automatic fault locating system. This makes for quick identification of fault location. One example shown in Figure 3.5.5., measures the two propagation times for surge current signals, generated at the fault point, to travel to the ends of the cable.

G. Capacitance Measurement Method

This method requires that the capacitance from one end of the faulted cable to ground be measured and compared in terms of distance with the capacitance of an unfaulted conductor in the same cable.[1] Almost any AC capacitance bridge is suitable, provided it measures capacitance to ground. In the absence of an AC bridge, a direct-reading capacitance meter can be employed.

In the presence of simultaneous ground faults, simultaneous open conductors or a low resistance path between broken conductors, it is necessary to connect the far end of the faulted phase to ground, either directly or via a resistance equal in value to the surge impedance of the cable. The measurement indicated by a direct-reading capacitance meter may contain a substantial error when the fault resistance is less than 1 MΩ.

H. Selection of Measuring Methods Suitable for Particular Types of Faults

Table 3.5.2. summarizes various fault-locating methods and comments on their appropriateness for different tasks.

I. Tracer Methods

Tracer methods are those which involve the placing of an electrical signal on the faulted conductor at one or both terminations and then tracing it along the cable length; faults are detected by a change in signal characteristics. Low-frequency currents and DC or impulse voltages are applied to the faulted cable and the phenomenon occurring at the point of failure can be surveyed by the magnetic, electromagnetic, acoustical, and electrical pick-up methods. Table 3.5.3. classifies the method according to detecting technique and comments on the advantages and disadvantages of each technique.

Table 3.5.2.
SUITABLE SELECTION OF FAULT LOCATORS

Series Resistance \ Fault	Fault Grounding Resistance and Discharge Voltage		
	Low Resistance(<kΩ) Path to Ground	High Resistance (>MΩ) Path to Ground	
		Discahrge Inception Voltage 4~5 kV	Discahrge Inception Voltage >10 kΩ
Single Line Grounded	LV Murray Loop Method / HV Murray Loop Method	LV Murray Loop (after burning) / HV Murray Loop Method / Discharge Radar Method	HV Murray Loop (after burning)* / Discahrge Radar Method
Two Lines Grounded — Parallel Line Available	"	"	"
Two Lines Grounded — No Parallel Line Available	LV Pulse Radar Method / Discharge Radar Method	LV Pulse Radar (after burning) / Discharge Radar Method	Discharge Radar Method
Two Lines Short-Circuited	LV Murray Loop Method	HV Murray Loop Method	HV Murray Loop (after burning)*
Three Lines Short-Circuited	LV Pulse Radar Method / Discharge Radar Method	LV Pulse Radar (after burning) / Discharge Radar Method	Discharge Radar Method
Cable Disconnected	AC Bridge Method / LV Pulse Radar Method	AC Bridge Method / Direct-Read Capacitance Meter / LV Pulse Radar Method	AC Bridge Method / Direct-Read Capacitance Meter / LV Pulse Radar Method

* It is often imposible or requires extremely long time to burn a point of failure open.

The measurement system comprising an HV pulse generator, an LV pulse radar measuring instrument, and a time-interval accurate to within one nanosecond seems to be applicable to almost all the cases described above.

Table 3.5.3.

SEVERAL TRACER METHODS TO DETECT FAULT POINTS

Name of Method	Measuring Principle	Advantages	Disadvantage
Search Coil Method	Low-frequency current is flowed between conductor and shield or ground, and resulting magnetic field is detected by a search coil.	1. Simple instrument 2. Easy to operate 3. Effective for Disconnection Fault 4. Effective to detect a defect of corrosion protective converings.	1. Low sensitivity in case of conductor-shield connection 2. Hardly applicable in case of high fault resistance (>1 kΩ) 3. Needs much experience of fault location
Discharge Detection Method	High voltage applied between conductor and shield produces discharges at a point of failure, and resulting electromagnetic wave is detected.	1. High Sensitivity. Applicable for a cable buried up to 2 m below the ground surface. 2. Effective for high resistance path to ground.	1. Rather complicated instrument 2. Difficult to operate 3. Needs some experience of fault location
Acoustic Method	Like the discharge detection method, discharges produced at a point of failure are detected acoustically by a microphone.	1. Suitable for cables installed in a tunnel or pipe. 2. Applicable for high resistance path to ground, too.	1. Low sensitivity. Difficult to detect in case a cable is buried 0.5 m below the surface. 2. Complicated instrument
Electric Potential Method	Potential induced by leakage current is measured by an electrometer with high input impedence.	1. Simple instrument 2. Easy to operate 3. High sensitivity 4. Applicable to detect a defect of corrosion protective coverings, too.	1. Ineffective to a shielded cable in general.

REFERENCES

1. **Anon.,** *Underground Systems Reference Book,* Edison Electric Institute, New York, 1957, chap. 11.
2. **Iizuka, K., Ed.,** *Power Cable Technology Handbook,* (in Japanese), Denki-Shoin, Tokyo, 1974, 1.
3. **Anon.,** *Insulation Testing Handbook,* (in Japanese), Institute Electrical Engineers of Japan, Tokyo, 1971, 1.
4. **Anon,** *Specifications for Impregnated-Paper-Insulated Low-Pressure Oil-Filled Cable,* 6th ed., Assoc. of Edison Illum. Companies, New York No. 4-69, 1969, 1.
5. **Anon.,** *Specifications for Impregnated-Paper-Insulated Cable — High-Pressure Pipe-Type,* 2nd ed., Assoc. of Edison Illum. Companies, New York, No. 2-67, 1967, 1.
6. **Anon.,** *Specifications for Polyethylene and Crosslinked Polyethylene Insulated Shielded Power Cables Rated 5 through 69 kV,* 5th ed., Assoc. of Edison Illum. Companies, New York, No. 5-75, 1975, 1.
7. IEEJ Committee Report, *Recommended High Voltage Testing for OF Cables,* (in Japanese), Institute of Electrical Engineers of Japan, Tokyo, Tech. Rep. No. 63, 1964, 24.
8. **Burrel, R. W. and Young, F. S .,** EEI Manufacturers 500/550 kV cable research project — Waltz Mill testing facility, *IEEE Trans. Power App Syst.,* 90(1), 180, 1971.
9. **Anon.,** Study of Transmission and Distribution Laboratory Facilities, NTIS PB-245 144, National Technical Information Service, Springfield, Va., 1975.
10. **Cavalli, M. and Mosca, W.,** A New Plant for Life-Tests on UHV Power Cables, Proc. Int. H. V. Symp., Zürich, 1975, 674.
11. IEEJ Committee Report, *Recommended High Voltage Test Methods for 11—77 kV XLPE Insulated Cables,* (in Japanese), Institute of Electrical Engineers of Japan, Tokyo, Tech. Rep. No. I-112, 1975, 1.
12. **Lee, R. H.,** *New Developments in Cable System Testing,* IEEE Conf. Rec. 10th Annual Meet. Ind. Appl. Soc., IEEE Piscataway, New Jersey, 1975, 820.
13. **Tanaka, T., Fukuda, T., and Suzuki, S.,** Water tree formation and lifetime estimation in 3.3 kV and 6.6 kV XLPE and PE power cables, *IEEE Trans. Power Appor. Syst.,* 95(6), 1892, 1976.
14. **Mason, J. H.,** Discharge detection and measurements, *Proc. Inst. Electr. Eng.,* 112, 1407, 1965.
15. **Kreuger, F. H.,** *Discharge Detection in High Voltage Equipment,* A Heywood Book, Temple Press Books, London, 1964, 1.
16. **Arnold, A.,** An Introduction to Partial Discharge Detection in Cables, Lecture Note, Brighton College of Technology, Brighton, England.
17. **Charters, J. S. T., Rooldset, S. A., and Salvage, B.,** Current pulse due to discharges in gaseous cavity in solid dielectrics, *Electr. Lett.,* 6(18), 569, 1970.
18. **Anon.,** *Guide for Test Procedure for Determining Partial - Discharge Extinction Level,* ICPEA T-24-380, Insulated Power Cable Eng. Assoc., Belmont, Mass., 1974.
19. **Black, I. A.,** *A Pulse Discrimination System for Discharge Detection Measurements on Equipment Operating in A Power System,* IEEE Conf. Publ. 94, Part 1, Piscataway, N.J., 1973, 1.
20. **Wilson, A.,** *The Application of Correlation Analysis of Partial Discharge Measurements,* IEEE, Piscataway, New Jersey, Conf. Publ. 94, Part 1, 1973, 8.
21. **Wilson, A.,** Discharge detection under noisy conditions, *Proc. Inst. Electr. Eng.,* 121(9), 993, 1974.
22. **Bartnikas, R.,** Use of a mutichannel analyzer for corona pulse height distribution measurements on cables and other electrical apparatus, *IEEE Trans. Instrum. Meas.,* 22(4), 403, 1973.
23. **Austin, J. and James, R. E.,** One-line digital computer system for measurement of partial discharges in insulation structures, *IEEE Trans. Electr. Insul.,* 11(4), 129, 1976.
24. **Merrel, E. J.,** Freezing oil-type pipe cables, *J. Am. Inst. Electr. Eng.,* 74(III), 1023, 1955.
25. **Genschke, G. P., May, K. R., and Thomas, J. A.,** Fault locator speeds restoration, *Electr. World,* January 1975, p. 42.
26. **Chen, C. S., Roemer, L. E., and Grumbach, R. S.,** *Power Cable Diagnostics Using Ceptrum Processing of Time Domain Reflectometry,* IEEE PES Summer Meeting A76 365-7, IEEE, Piscataway, New Jersey, 1976, 1.

APPENDIX

TABLE OF SCALE CONVERSION

Quanity						SI unit
Length	1	mile	1609	km	1.61×10^3	m
	1	ft	30.48	cm	3.05×10^{-1}	m
	1	in.	25.4	mm	2.54×10^{-2}	m
	1 mil = 10^{-3}	in.	25.4	μm	2.54×10^{-5}	m
Area	1	acre	4046.9	m²	4.05×10^3	m²
	1	in.²	645	mm²	6.45×10^{-4}	m²
	1000	kcmil	506.7	mm²	5.07×10^{-4}	m²
Volume	1	gal	3785.43	cm³	3.79×10^{-3}	m³
	1 British gallon = 1.2 gal	gal	4542.52	cm³	4.54×10^{-3}	m³
Weight	1	lb	453.6	g	4.54×10^{-1}	kg
Density	1	lb/ft³	0.0160	g/cm³	1.60×10	kg/m³
Electric field	100	V/mil	3.94	kV/mm, MV/m	3.94×10^6	V/m
Pressure	100	psi	7.03	kg/cm²		
	14.22	psi	1 atm, 760 mmHg	kg/cm²	0.0981	MPa
	145	psi	10.20	kg/cm²	1	MN/m²
Energy and heat	1	Btu	1054.866	J	1.05×10^3	J
	1	cal	4.186	J	4.186	J
Flow rate	100	gpm	6.31 ℓ/sec, 22.7	m³/hr	6.31×10^{-3}	m³/sec
Thermal conductivity	1 Btu/hr × ft	°F	1.7305×10^{-2}	W/cm°C	1.73	W/m K
	1 Btu × in./hr × ft	°F	1.4420×10^{-3}	W/cm°C	1.44×10^{-1}	W/m K
Temperature	°F	°C = (5/9) (°F − 32)			K = °C + 273.15	

VISCOSITY CONVERSION TABLE

Poise   = c.g.s. unit of absolute viscosity = $\dfrac{g}{sec \times cm}$

Stoke   = c.g.s. unit of kinematic viscosity = $\dfrac{g}{sec \times cm \times density(t°F)}$

Centipoise   = 0.01 poise
Centistoke   = 0.01 stoke
Centipoises = Centistokes × density (at given temperature)

To convert poises to $\dfrac{lb}{sec \times ft}$ or $\dfrac{lb}{hr \times ft}$ multiply by 0.0672 or 242, respectively.

| | Saybolt seconds at | | | Redwood seconds at | | | Engler degrees at |
Centistokes	100°F	130°F	210°F	70°F	140°F	200°F	all temps.
2.0	32.6	32.7	32.8	30.2	31.0	31.2	1.14
3.0	36.0	36.1	36.3	32.7	33.5	33.7	1.22
4.0	39.1	39.2	39.4	35.3	36.0	36.2	1.31
5.0	42.3	42.4	42.6	37.9	38.5	38.9	1.40
6.0	45.5	45.6	45.8	40.5	41.0	41.5	1.48
7.0	48.7	48.8	49.0	43.2	43.7	44.2	1.56
8.0	52.0	52.1	52.4	46.0	46.4	46.9	1.65
9.0	55.4	55.5	55.8	48.9	49.1	49.7	1.75
10.0	58.8	58.9	59.2	51.7	52.0	52.6	1.84
11.0	62.3	62.4	62.7	54.8	55.0	55.6	1.93
12.0	65.9	66.0	66.4	57.9	58.1	58.8	2.02
14.0	73.4	73.5	73.9	64.4	64.6	65.3	2.22
16.0	81.1	81.3	81.7	71.0	71.4	72.2	2.43
18.0	89.2	89.4	89.8	77.9	78.5	79.4	2.64
20.0	97.5	97.7	98.2	85.0	85.8	86.9	2.87
22.0	106.0	106.2	106.7	92.4	93.3	94.5	3.10
24.0	114.6	114.8	115.4	99.9	100.9	102.2	3.34
26.0	123.3	123.5	124.2	107.5	108.6	110.0	3.58
28.0	132.1	132.4	133.0	115.3	116.5	118.0	3.82
30.0	140.9	141.2	141.9	123.1	124.4	126.0	4.07
32.0	149.7	150.0	150.8	131.0	132.3	134.1	4.32
34.0	158.7	159.0	159.8	138.9	140.2	142.2	4.57
36.0	167.7	168.0	168.9	146.9	148.2	150.3	4.83
38.0	176.7	177.0	177.9	155.0	156.2	158.3	5.08
40.0	185.7	186.0	187.0	163.0	164.3	166.7	5.34
42.0	194.7	195.1	196.1	171.0	172.3	175.0	5.59
44.0	203.8	204.2	205.2	179.1	180.4	183.3	5.85
46.0	213.0	213.4	214.5	187.1	188.5	191.7	6.11
48.0	222.2	222.6	223.8	195.2	196.6	200.0	6.37
50.0	231.4	231.8	233.0	203.3	204.7	208.3	6.63
60.0	277.4	277.9	279.3	243.5	245.3	250.0	7.90
70.0	323.4	324.0	325.7	283.9	286.0	291.7	9.21
80.0	369.6	370.3	372.2	323.9	326.6	333.4	10.53
90.0	415.8	416.6	418.7	364.4	367.4	375.0	11.84
100.0[a]	462.0	462.9	465.2	404.9	408.2	416.7	13.16

[a] At higher values use the same ratio as above for 100 cSt: e.g., 100 cSt = 110 × 4.620 Saybolt sec at 100°F.

To obtain the Saybolt Universal viscosity equivalent to a kinematic viscosity determined at $t°F$, multiply the equivalent Saybolt Universal viscosity at 100°F by 1 + $(t - 100)$ 0.000064; e.g., 10 cSt at 10°F are equivalent to 58 × 1.0070, or 59.2 Saybolt Universal seconds at 210°F.

From Weast, R., *CRC Handbook of Chemistry and Physics*, 56th ed., CRC Press, Boca Raton, Fla., 1975-76, F-47.

INDEX

Dosage, 116
Double bonds, 93—94, 116
Drop in pressure, 186
Dry ice, 186
Drying, 36
Duplex cable, 7, 54

E

Echo pulses, 191, 192
Eddy current losses, 22, 28
Effective coefficient of ionization, 80, 81
EHV, see Extra-high voltage
Einstein's relation, 104
Elastic collisions, 82
Elastic limit, 109
Electrets, 129
Electrical breakdown characteristics, 66—86
Electrical design, 7—13
Electrical energy, 1—2
    growth in consumption of, 1
Electrical failure, 186
Electrical resistance, 12
Electrical stress
    approximate formulas for, 143—146
    maximum, 7
    uniform temperature distribution conditions and,
        136—143
Electrical tests, 169—172
Electrical trees, 87, 96, 106
Electric field
    cylindrical geometry and, 140
    Hippel, 67
    plane-parallel geometry and, 137—140
    protrusions and, 140—143
    space charge-modified, 143
Electric field distribution, 128
    homocharge case and, 136—137
Electrochemical trees, 104
Electrode systems, 141
Electrolytes, 103
Electrolytic corrosion, 62
Electromechanical breakdown, 74
Electron affinity, 84
    dielectric, 132
Electron avalanche, 66—70, 84, 87—90
Electron bubble, 84
Electronegative gases, 80
    compressed, 81
Electron emission, 85
Electronic homocharge, 132—136
Electronic thermal breakdown, 74, 75
Electron injection, 130—131, 146
Electron multiplication factor, 89
Electrons
    high-energy, 103
    hopping, 74
    hot, 90
    reservoir of, 135
Electrostatic screens, 49

Electrostrictive force, 74, 80, 104, 106, 109
Elongation, 112
Emanueli units, 37
Emergency operating temperatures, 13
Emission
    current limited by, 136, 140
    electron, 85
    field, 76—77, 90—91, 130
    Schottky, 130—132, 136, 148
Endothermic reactions, 94
Energy
    activation, 114
    discharge, 176
    electrical, see Electrical energy
    forecasting of, 1
    geothermal, 2
    solar, 2
    tidal, 2
    wind, 2
Epoxy coating, 64
Epoxy resins, 87
EPR, see Ethylene-propylene rubber
Equi-pressure method, 188
Equivalent time, 114
Estimation formula of skin and proximity effects,
        28—32
Ethyl acrylate, 60
Ethylene, 47
Ethylene copolymers, 60
Ethylene-ethyl acrylate copolymer, 60
Ethylene-propylene copolymer, 4, 60
Ethylene-propylene-dien terpolymer, 4
Ethylene-propylene rubber (EPR), 10, 44, 47, 104
Ethylene-vinyl acetate copolymer, 60
Ethylvinyl acetate (EVA), 7
Ethylvinyl acetate (EVA) copolymer, 8
EVA, see Ethylvinyl acetate
Excitation, 82
Exhaustion layer, 132
Expansion, 104
Exponential distribution, 117, 121
Expulsion test for OF cable, 159
External interference, 178
Extraction
    charge, 99
    solvent, 167
Extra-high voltage (EHV), 3
Extreme distribution, 117
Extruded cable, 7, 10, 11
    testing of, 164—168
Extruded sheath, 159

F

Factory tests, 157—162, 164—167
    on OF and POF cables, 159—162
Failure
    cumulative rate of, 127
    electrical, 186
    initial, 121

P